Environmental Efficiency, Innovation and Economic Performances

Eco-innovation is becoming a conceptual reference point for many regional and international public policies and management strategies. This field of research has been focusing on how environmental innovation is particularly related to the intensity of emissions and economic performance. There are two reasons for this growing interest. The first is that environmental performance is one of the main economic policy goals of European countries, thanks to its relevance to the Lisbon Strategy and the Göteborg priorities for sustainable development. The second, which is partly linked to the first, is related to the growing impact of environmental regulation on private sector activity in many European countries.

This volume brings together microeconomics studies on firms' eco-innovation and economic performance, both in the industrial and service sector, with a sector-based perspective rooted mainly in the exploitation of NAMEA data at regional level, and finally with a macroeconomic analysis of the environment, income and welfare.

This collection brings together the best of recent research in the interlinked areas of eco-innovation and income-environment relationships studies, and in its entirety is an excellent source of knowledge for postgraduates, researchers and students of Environmental and Ecological Economics alike. As well as fully developing the theoretical aspects of its topics, these essays are also strongly policy-oriented and will be of interest to anyone seeking information on an applied perspective.

Massimiliano Mazzanti is Associate Professor in Environmental Economics at the University of Ferrara, Italy.

Anna Montini is Assistant Professor in Economics and Environmental Economics at the University of Bologna, Italy.

Routledge Explorations in Environmental Economics

Edited by Nick Hanley
University of Stirling, UK

1 **Greenhouse Economics**
Value and ethics
Clive L. Spash

2 **Oil Wealth and the Fate of Tropical Rainforests**
Sven Wunder

3 **The Economics of Climate Change**
Edited by Anthony D. Owen and Nick Hanley

4 **Alternatives for Environmental Valuation**
Edited by Michael Getzner, Clive Spash and Sigrid Stagl

5 **Environmental Sustainability**
A consumption approach
Raghbendra Jha and K.V. Bhanu Murthy

6 **Cost-Effective Control of Urban Smog**
The significance of the Chicago Cap-and-Trade approach
Richard F. Kosobud, Houston H. Stokes, Carol D. Tallarico and Brian L. Scott

7 **Ecological Economics and Industrial Ecology**
Jakub Kronenberg

8 **Environmental Economics, Experimental Methods**
Edited by Todd L. Cherry, Stephan Kroll and Jason F. Shogren

9 **Game Theory and Policy Making in Natural Resources and the Environment**
Edited by Ariel Dinar, José Albiac and Joaquín Sánchez-Soriano

10 **Arctic Oil and Gas**
Sustainability at risk?
Edited by Aslaug Mikkelsen and Oluf Langhelle

11 **Agrobiodiversity, Conservation and Economic Development**
Edited by Andreas Kontoleon, Unai Pascual and Melinda Smale

12 **Renewable Energy from Forest Resources in the United States**
Edited by Barry D. Solomon and Valeria A. Luzadis

13 **Modeling Environment-Improving Technological Innovations under Uncertainty**
Alexander A. Golub and Anil Markandya

Environmental Efficiency, Innovation and Economic Performances

Edited by
Massimiliano Mazzanti
and Anna Montini

Routledge
Taylor & Francis Group
LONDON AND NEW YORK

First published 2010
by Routledge
2 Park Square, Milton Park, Abingdon, Oxon, OX14 4RN

Simultaneously published in the USA and Canada
by Routledge
270 Madison Avenue, New York, NY 10016

Routledge is an imprint of the Taylor & Francis Group, an informa business

Typeset in Times New Roman by
HWA Text and Date Management, London

British Library Cataloguing in Publication Data
A catalog record for this book is available from the British Library

Library of Congress Cataloging in Publication Data
Environmental efficiency, innovation and economic performances / edited by Massimiliano Mazzanti, Anna Montini.
p. cm.
Includes bibliographical references and index.
1. Industrial management – Environmental aspects. 2. Industries – Environmental aspects. 3. Technological innovations – Environmental aspects. 4. Organizational change – Environmental aspects. I. Mazzanti, Massimiliano. II. Montini, Anna.

HD30.255.E5777 2010
658.4′083–dc22 2009049512

ISBN: 978-0-415-47852-6 (hbk)
ISBN: 978-0-203-85041-1 (ebk)

To Roberto

Man's mind cannot grasp the causes of events in their completeness, but the desire to find those causes is implanted in man's soul. And without considering the multiplicity and complexity of the conditions any one of which taken separately may seem to be the cause, he snatches at the first approximation to a cause that seems to him intelligible and says: 'This is the cause!'

Tolstoy, *War and Peace*, Planet pdf, page 2324

Contents

Figures

Tables

Contributors

Emilie Alberola is Economist at the Mission Climat de la Caisse des dépôts, France and has a PhD from the University of Paris I.

Angelo Antoci is Full Professor of Mathematics for Economics at the University of Sassari. He was previously Assistant Professor at the University of Florence (1993–98) and Associate Professor at the University of Sassari (1998–2005).

Davide Antonioli is a Research Assistant and Lecturer in Labour Economics at the University of Ferrara. He received his MSc in Economics from the University of Bristol and was awarded a PhD in Economics by the University of Bologna.

Simone Borghesi has a MSc in Economics from University College London and a PhD from the European University Institute in Florence. He is Assistant Professor and Lecturer at the University of Siena. He has worked at the International Monetary Fund (1998), the Fondazione ENI Enrico Mattei (1999) and the University of Pescara (2004–08).

Giulio Cainelli is Full Professor of Economics at the University of Padua, and Senior Research Fellow at the CERIS-DSE in Milan.

Maurizio Canavari holds a Laurea degree in Agricultural Sciences from Bologna University and a Doctorate in Appraisal and Land Economics from the University od Padua. He is currently Associate Professor of Agricultural Economics and Appraisal at the Alma Mater Studiorum University of Bologna.

Nicola Cantore has PhDs in Economics from the Università Cattolica del Sacro Cuore, Milan, and in Environmental Economics and Management from the University of York. He was Senior Research Fellow at the University of Bologna, Faculty of Agriculture, Department of Agricultural Economics and Engineering. He is currently Research Fellow at the Overseas Development Institute in London.

Julien Chevallier is Assistant Professor in Economics at the Université Paris Dauphine. He is member of the Centre de Géopolitique de l'Energie et des Matières Premières (CGEMP) and the Laboratoire d'Economie de Dauphine (LEDa). He is also Visiting Researcher with EconomiX-CNRS at the Université

Paris Ouest Nanterre La Défense and the Grantham Institute for Climate Change at Imperial College London.

Benoît Chèze is a PhD student at the University Paris-X, Nanterre.

Valeria Costantini is Assistant Professor and Lecturer in the Department of Economics of the University of Roma Tre, and also lectures in Urban Economics. She holds a PhD in Institutions, Environment and Development Economics from the University of Roma Tre. She was formerly an Economist at the Italian National Research Institute for New Technologies, Energy and Environment (ENEA).

Francesco Crespi is Assistant Professor and Lecturer in the Department of Economics of the University of Roma Tre, and Research Associate at the Bureau of Research on Innovation, Complexity and Knowledge (BRICK), Collegio Carlo Alberto. He has a PhD in Economics. He was formerly an economist in the Italian Treasury and was a Marie Curie Fellow at the Institute of Innovation Research, University of Manchester.

Aldo Femia is Senior Researcher at the Italian National Statistical Institute (ISTAT) and has worked at the Wuppertal Institute, and the OECD in Paris. He was awarded an MSc in Economics and Econometrics from the University of Southampton, and holds a PhD in Political Economy from University Federico II Naples

Michael Getzner is Associate Professor at the Department of Economics, Klagenfurt University. He has worked in the fields of Environmental and Ecological Economics, Public Finance, and Economic and Regional Policy. He is Director of the Master's course in Management of Protected Areas.

Giovanni Marin achieved his postgraduate degree from the Faculty of Economics at the University of Ferrara and is currently a doctoral student at the IMT in Lucca.

Renato Marra Campanale is a researcher at the Italian Environmental Protection Institute. He worked for a time at the OECD and cooperates with ISTAT on environmental accounting. He holds a Master's degree in Environmental Policy from Pavia University, Italy.

Massimiliano Mazzanti is a Lecturer in Environmental Economics and Associate Professor at the University of Ferrara. For many years he has been a Research Fellow with CERIS-CNR, an institute of the National Research Council in Milan. He was awarded an MSc in Environmental and Resource Economics from the University College of London and a PhD in Institutions, Environment and Development Economics from the University of Rome Tre. Email: ma.maz@iol.it.

Anna Montini is Assistant Professor in Economics and Lecturer in Economics and Environmental Economics at the University of Bologna, Italy. She is a

Research Fellow at the National Research Council in Milan (CERIS-DSE). She holds a PhD in Political Economics from the University of Siena. Email: anna.montini@unibo.it

Antonio Musolesi is a researcher at INRA Dijon, France. He is currently a visiting researcher at the Department of Geography and Environment at the London School of Economics and Political Science. He has a PhD in Economics from the University of Nice-Sophia Antipolis and an MSc in Statistics and Econometrics from the University of Toulouse.

Serena Pontoglio is a Research Fellow at CERIS-CNR (National Council of Research, Institute for Economic Research on Firms and Growth) in Milan. She completed her PhD on the Implications of the European Emissions Trading Scheme on the Eco-innovation Process.

Annalisa Zezza is a Senior Researcher in Agricultural Policy and Agri-environment at the Italian National Institute of Agricultural Economics (INEA). She has a PhD in Agricultural Economics from the University of Naples. From April 1998 to March 2004 she was General Director of the INEA and a member of the board of the Italian Society of Agricultural Economics.

Roberto Zoboli is Full Professor in Political Economy at the Catholic University of Milan and for many years was Research Director at CERIS-DSE in Milan. He lectures on the international Masters' courses at the Catholic University and since 2003 has coordinated numerous projects on environmental policy under the umbrella of the European Environmental Agency.

Preface

This book collects several pieces originating from various recent research projects with participation from CERIS DSE Institute of the National Research Council, the Department of Economics of the University of Bologna and the Department of Economics Institutions and Territory of the University of Ferrara. Over the years, colleagues have developed research streams on drivers of environmental innovation and effects on economic performances, analyses of delinking regarding the income–environment dynamic relationship, and assessment of the environmental Kuznets curve hypothesis.

This volume brings together micro-economic studies (Part I, Chapters 1–4), a sector-based perspective, rooted mainly in the exploitation of NAMEA data (Part II, Chapters 5–9), and a macroeconomic analysis of the environment, income and welfare (Part III, Chapters 10–13). These works offer value added and originality – primarily from an applied perspective, although theoretical aspects are not neglected and are discussed in length or fully developed (Part III). Most of the chapters are strongly policy oriented. For all these reasons we believe that this book may be attractive to a wide audience, extending beyond academia.

On the one hand, this book collects past and ongoing research undertaken by members of the aforementioned institutions; on the other hand, it links those consolidated research efforts with contributions from promising young scholars, friends and colleagues, who have had extensive involvement with the issues discussed. Thus, it provides the state of the art in our recent research and a foundation for possible future research at national and EU levels: a challenge for future extension research from further amalgamation of ideas, issues and research projects.

Acknowledgements

We want to acknowledge the financial support received by various institutions and projects over the years. Massimiliano Mazzanti acknowledges funding received by the local project 2008 on Innovation, Industrial Relations and Productivity in Local Systems, financed by the University of Ferrara, the 2005–2007 research programme on Innovative Dynamics in the Knowledge-based Economy, funded by the University of Ferrara budget for projects of local interests (60 per cent), and PRIN-MIUR 2003–2004 on Structural Dynamics: Firms, Organizations, Institutions. He is grateful to the CGIL, especially the trade union branch located in Reggio Emilia, for its support. Some data are derived from surveys funded within the project "Infrastructures, Competitiveness and Levels of Government: Knowledge and Development of the New Economy" (PRIN 2001–2002). Massimiliano Mazzanti also would like to thank the many postgraduate students he supervised at the University of Ferrara over the past years, including Chiara Marinelli and Giovanni Marin. Their contribution to the overall work has been invaluable. Anna Montini acknowledges the financial support received by the PRIN project "Emerging economic regional powers and local systems of production: new threats or new opportunities?" We thank Cynthia Little for support regarding the English revision of final versions.

The authors also acknowledge the ongoing support of CERIS-CNR. We want to express our intellectual debt to Roberto Zoboli, a friend, and an outstanding environmental economist and colleague. Most of the works collected in this volume are dependent on his crucial contribution or derive from the intellectual stimulus and support that he has provided over the years. The formal and informal discussions about environmental economics and environmental policy issues were and are invaluable. To an extent this book is dedicated to him and his contribution, but also is pointing the way to new intellectual challenges.

Abbreviations

2SLS	two-stage least squares
ADF	Augmented Dickey–Fuller
AIDA	Analisi Informatizzata Delle Aziende (Informative Analysis of Firms)
ANBERD	Analytical Business Enterprise Expenditure on Research and Development
BAU	business as usual
CAC	command and control
CAFE	Clean Air for Europe
CAFTA	Central American Free Trade Agreement
CBI	Caribbean Basin Initiative
CCS	carbon capture and storage
CDIAC	Carbon Dioxide Information Analysis Centre
CDM	clean development mechanism
CHN	China
CHP	combined heat and power
CIS	Community Innovation Survey
CKC	Carbon Kuznets Curve
CO_2	carbon dioxide
COMTRADE	Commodity Trade Statistics Database
CSR	corporate social responsibility
DK	Driscoll–Kraay
DOLS	dynamic ordinary least squares
DSUR	dynamic seemingly unrelated regressions
EC	European Commission
EE	Eastern Europe(an)
EFTA	European Free Trade Agreement
EI	eco-innovation
EIs	economic instruments
EKC	Environmental Kuznets Curve
EMS	Environmental Management Systems
ENEA	Ente Nazionale Energia Ambiente (Italian National Institute for New Technologies, Energy and Environment)
ETS	Emissions Trading Scheme

EU	European Union
EU ETS	European Union Emissions Trading Scheme
EUA	European Union Allowance
FEM	Fixed Effects Model
FTEJ	full time equivalent jobs
GARCH	general autoregressive conditional heteroskedasticity
GDP	gross domestic product
GHG	greenhouse gases
GK	Global Kyoto
GLS	generalized least squares
GNP	gross national product
GSI	Institute for Sustainable Development's Global Subsidies Initiative
GSP	Generalized System of Preferences
GWP	global warming potential
HCSE	heteroskedasticity-consistent standard errors
HDI	Human Development Index
HPWP	high-performance work practices
HRM	human resource management
ID	industrial districts
IEA	International Energy Agency
IPAT	impact population affluence technology
IPPC	integrated pollution prevention control
IR	industrial relations
ISIC	International Standard Industrial Classification
ISO	International Organization for Standardization
ISPRA	Istituto Superiore per la Protezione e la Ricerca Ambientale
JI	joint implementation
LIC	low income countries
LMIC	lower middle income countries
LPS	local production systems
LSD	least square dummy
MBI	market-based instruments
MERCOSUR	Mercado Comùn del Sur
MFN	most-favoured nation
MG	mean group
MIC	middle income countries
ML	maximum likelihood
NACE	Nomenclature statistique des activités économiques dans la Communauté européenne
NAFTA	North American Free Trade Agreement
NAMEA	National Accounting Matrix for Environmental Accounts
NAP	National Allocation Plan
NBP	NO_x Budget Trading Programme
NO_x	nitrogen oxide

NWE	New Welfare Economics
OECD	Organization for Economic Cooperation and Development
OHI	other high income
OLS	ordinary least squares
PAE	potential acid equivalents
PIR	pollution–income relationship
PMG	Pooled Mean Group
PP	Phillips–Perron
PPP	purchasing power parity
R&D	research and development
RAMEA	Regional Accounting Matrix for Environmental Accounts
RECLAIM	Regional Clean Air Incentives Market
REM	aandom effects model
RICE	regional integrated climate economy model
SEA	system of the enterprise accounts
SME	small and medium enterprises
SNAP	Selected Nomenclature for Sources of Air Pollution
SO_x	sulphur oxides
STIRPAT	Stochastic Impacts by Regressions on Population, Affluence and Technology
SUR	seemingly unrelated regressions
SWF	social welfare function
TJ	Terajoule
Tobit	Tobin Probit Model
TOFP	tropospheric ozone forming potentials
TP	turning point
UK	United Kingdom
UNCTAD	United Nations Conference on Trade and Development
UNDP	United Nations Development Programme
UN-TRAINS	United Nations Trade Analysis and Information System
US	United States
USA	United States of America
VAR	vector autoregression
WDI	world development indicators
WSP	world social planner
WTO	World Trade Organization

Introduction

We provide a short conceptual sketch of the scientific framework in which the book is embedded, of which eco-innovation (EI) is probably the corner stone. EI is becoming a conceptual reference point for many regional and international public policies and management strategies. Over the first years of the twenty-first century, this field of research has been focusing on environmental innovations, particularly related to the intensity of emissions, and economic performance and efficiency. There are two reasons for this growing interest. The first is that environmental performance is one of the main economic policy goals of European countries thanks to its relevance in the Lisbon Strategy and the Göteborg priorities for sustainable development. The second, which is partly linked to the first, is related to the growing impact of environmental regulation on private sector activity in many European countries.

One of the most recent definitions of eco-innovation characterizes it as the production, application or use of a product, service, production process or management system that is new to the firm adopting or developing it, and which implies a reduction in environmental impact and resource use (including energy) throughout its life-cycle. This definition includes innovations whose environmental effects are not intentional. A relevant distinction can be made between end-of-pipe technologies and clean technologies integrated in production processes. In addition to these two categories, there are eco-products, eco-processes, eco-management systems, and broader environmental innovation systems.

Studies on environmental innovation and its relationship with socio-economic performance originate within three analytical fields within the framework of environmental economics and management/innovation studies. Below we outline the main research fields.

One research field regards the efficacy and dynamic efficiency of environmental policy tools as objectives to be evaluated along with static efficiency (Requate and Unhold 2003; Requate 2005). The analysis of efficacy, linked to the specific goal of ex-post evaluation and analysis of the effects on innovative paths induced by policies (Mazzanti and Zoboli 2006; Kemp 1997), is complemented by the analysis of dynamic efficiency (Milliman and Prince 1989; Hahn and Stavins 1994; Van den Bergh 2007). There is another research field related to Porter's hypotheses on the links between environmental regulation and innovative strategies on the one hand, and between the performance/competitiveness of firms and Corporate Social Responsibility (CSR) strategies on the other. On the heels of the work of Porter (Porter and van der Linde 1995), many authors have tackled the issues of the effects on manufacturing sectors most vulnerable to competition and regulations (Jaffe *et al.* 1995), and the motivations of CSR, whether explainable by profit-only or by the production of mixed public goods (Reinhardt *et al.* 2008; Rubbelke and Markandya 2008). These conceptual fields of research lead us to empirical research that focuses on the drivers of environmental innovation and its effects on performance, which is a core part of this book.

The literature, much of it originating in the United States (US), features studies on the drivers of innovation, focusing on formal innovation (patents), which is more relevant in this context, and on the internal drivers of innovation, such as R&D spending (Brunnermeier and Cohen 2003; Jaffe and Palmer 1997; Popp 2006, 2002; Johnstone *et al.* 2008). Although this literature is based on panel data, it is limited by its inclusion only of studies on certain innovative typologies, and analyses tailored to the US environmental and emissions policy framework. For these reasons, especially in Europe, the late 1990s saw the emergence of various applied studies based on the collection of micro-data through surveys. These surveys achieved high levels of detail and specificity with regard to the innovation variables, motivations and potential drivers analysed. These studies concentrate on the specific drivers of end-of-pipe and process integrated innovations (Frondel *et al.* 2008), eco-labelling processes (Johnstone and Labonne 2009), and the correlation between various technological and organizational innovative dynamics (Rennings *et al.* 2006). The analyses test both the relevance of drivers internal to firms, such as research and development (R&D), and of other external innovation strategy factors, such as networking, spillover effects, industrial relations and policy stringency, differentiated by sectors and regions (Horbach 2008). Nevertheless, there are ample opportunities for research in the European context, including research on green patents, especially through spatial analyses, which are rare in the field of environmental innovation. Along with data on attribution, patents also contain relational data.

The literature has achieved a high level of specificity with regard to the various micro-economic analyses. However, there is still room for research in fields that have not been fully explored, and there is a need for greater integration between analyses of the drivers and effects of environmental innovation. Effects have received less attention from the (European) empirical literature in the last decade, mostly due to difficulties in obtaining performance data that can be combined with environmental innovation data at the micro-economic level. Building upon the literature on the links between innovation and economic performance, recent works have looked at the effects of environmental innovation on economic performance.

The relationship between technological change and economic performance has long been the focus of economic studies, dating back to the fundamental contributions of the classical economists. This research tradition continued and grew with Schumpeter, who placed innovation at the core of his analyses following his seminal contribution in 1934, *The Theory of Economic Development*. In 1957, Solow furthered the analysis of the relationship between technological change and performance, addressing the issue within a theoretical framework based on the aggregate production function. Brock and Taylor (2005) develop a (new) Green Solow model that proposes a theoretical basis for the intrinsic empirical EKC framework.

Such works are still few in number, however, and do not constitute a body of literature as in the case of drivers (Earnhart and Lizal 2007; Ziegler *et al.* 2008). Overall, the evidence so far is neither robust nor univocal, and leaves room for new contributions that can identify empirically the direction and intensity of the relationship between environmental innovations and economic performance. The approach mainly adopted has been to consider environmental performance as a factor that tends to have a negative impact on private sector performance (Walley and Whitehead 1994), the idea being that in some industries, particularly those at the beginning of the production chain (such as extractive industries), environmental costs are a significant share of the added value of firms, thus negatively impacting their competitiveness. The idea that improved environmental performance can be a potential source of competitive advantage for the private sector has only recently come to the fore (Porter 1991; Schmidheiny 1992). The literature in this field identifies two main ways that this can happen: (i) the higher costs associated with polluting activities can encourage firms to acquire new technologies and production systems, which, in turn, entail lower production costs and thus higher productivity (Porter and Esty 1998); (ii) the first firms to adopt these new technologies can benefit from having made the first move, by selling these innovations to other firms.

Original value added may be derived from exploiting sector based data on environmental and economic accounts, which is challenging due to the paucity of data and potential need to integrate firm, sector and regional data (Cole *et al.* 2005). This meso level links microeconomic (firm based) considerations with macro perspectives, thus also creating a bridge analysis of economic and environmental performance, and innovation and policy factors, at firm level, and EKC analysis where income-environment relationships are assessed at sector, regional or country level, testing possible influences on the relationship with innovation, energy and policy drivers (Costantini and Crespi 2008).

The perspective of environmental accounting for the analysis of the relationships between the economic and environmental systems, especially in relation to satellite accounts such as NAMEA, is relatively recent (de Haan and Keuning 1996) and in part derives from the conceptual and applied deficits that have emerged during the setting up of green GDP or GNP measures as alternative measures of accounting. NAMEA provides a comprehensive and integrated picture of the economic system (production and value) in association with the environmental system (physical pressures such as emissions) by sector classification (Keuning *et al.* 1999). As a consequence of the limited temporal coverage, the first applications were in decomposition analyses, over periods of a few years (de Haan 2001).

Even less developed are studies that exploit RAMEA (Stauverman 2007), among which we highlight its recent applications to Lazio and Emilia Romagna (Mazzanti *et al.* 2007).

Thanks on the one hand to the completion of full panel datasets in some countries including Italy, and on the other hand to the increasing emphasis in the EU environment of delinking and policy effectiveness analyses (Jacobsen 2000), and the joint development of studies in the realms of EKC/IPAT and structural decomposition of environmental performance (EnP) drivers (Bruvoll and Medin 2003; Diakoulaki and Mandaraka 2007), a literature that investigates the stochastic relationships between environmental and socio-economic phenomena has become established (Martinez Zarzoso *et al.* 2007). A first set of studies uses NAMEA to exploit intrinsic sector and time variabilities (Mazzanti *et al.* 2008; Roca and Serrano 2007). The aim is to provide more circumscribed analyses with respect to the majority of cross-country studies in the prevailing (EKC) literature (Borghesi 2008). The causal links between measures of economic growth (labour productivity) and indicators of environmental (economic and technical) efficiency have been studied, and eventual non-linearity in dynamics, and absolute or relative delinking in local and global externalities has been confirmed (Mazzanti and Zoboli 2009).

EKC analysis addresses some of Impact-Population-Affluence-Technology (IPAT)-linked relationships, that is, between I and GDP (gross domestic product) or between T and GDP/P, by looking at the direct/indirect benefits and costs of growth in terms of environmental impact. Although such analyses may highlight empirical regularities that are of heuristic value, they do not provide direct economic explanations. The existence of an EKC deterministically could be misleading in suggesting that rapid growth towards high levels of GDP/P automatically produces greater environmental efficiency and, thus, that growth might be the best policy strategy to reduce environmental impact.

The link between environmental efficiency and economic performance is relevant in determining the drivers of T and hence the innovative objectives of the economic agents. Some of the recent theoretical and empirical literature focuses on assessment of the complementarities between the two performances. Prior to the paper by Porter and van der Linde (1995), the generally accepted view was that fulfilment of environmental regulation would reduce the competitiveness of the target sectors/firms. However, the so-called Porter hypothesis reveals the potential complementarities and (private) beneficial effects of properly designed environmental regulations. While in some cases environmental regulations are necessary, in others information about potentially promising paths to innovation is sufficient (e.g. early mover advantages). The adoption of eco-efficient processes in advance of competitors creates an economic advantage at the time when environmental regulation is introduced. The advantage comes from the inertia of competitors in adapting to the new environmental standards. Finally, it should be noted that, in general, ecological and economic innovations cannot be separated. So, ecological innovations might be a by-product of economic innovations.

We conclude with some policy-oriented reasoning. Taking account of national dynamics is highly relevant in reasoning around the underlying dynamics of emissions, policy implementation and policy effectiveness. The value of country-based delinking evidence is high. NAMEA structured studies could provide great value added in the policy arena as well as contributing to the EKC economic debate, wherein we highlight the increasing relevance of a focus on either a single country (exploiting geographical and sector heterogeneity) or on a homogenous set of countries or regions in order to provide policy relevant insights. NAMEA based analyses provide added value in EU research versus US applied environmental research (Huppes *et al.* 2005), and open up many directions for investigation.

These analyses represent further challenges with respect to the already mature literature on eco-innovation drivers and the effects of specifically tailored innovations, but which omit the strict environmental aims of innovation.

Massimiliano Mazzanti
Anna Montini
September 2009

References

Borghesi, S. (2008) 'From Hubbert to Kuznets: On the sustainability of the current energy system', *International Journal of Global Environmental Issues*, 8: 425–44.

Brock, W. and Taylor, M.S. (2005) 'Economic growth and the environment: A review of theory and empirics', in P. Aghion and S. Durlauf (eds) *Handbook of Economic Growth*, (vol 1B), Amsterdam: North Holland, pp. 1749–1821.

Brunnermeier, S. and Cohen, M. (2003) 'The determinants of environmental innovation in US manufacturing industries', *Journal of Environmental Economics and Management*, 45: 278–93.

Bruvoll, A. and Medin, H. (2003) 'Factors behind the environmental Kuznets curve. A decomposition of the changes in air pollution', *Environmental and Resource Economics*, 24: 27–48.

Cole, M., Elliott, R. and Shimamoto, K. (2005) 'Industrial characteristics, environmental regulations and air pollution: An analysis of the UK manufacturing sector', *Journal of Environmental Economics and Management*, 50: 131–43.

Costantini, V. and Crespi, F. (2008) 'Environmental regulation and the export dynamics of energy technologies', *Ecological Economics*, 66: 447–60.

De Haan, M. (2001) 'A structural decomposition analysis of pollution in the Netherlands', *Economic System Research*, 13: 181–96.

De Haan, M. and Keuning, S.J. (1996) 'Taking the environment into account: the NAMEA approach', *The Review of Income and Wealth*, 42: 131–48.

Diakoulaki, D. and Mandaraka, M. (2007) 'Decomposition analysis for assessing the progress in decoupling industrial growth from CO_2 emissions in the EU manufacturing sector', *Energy Economics*, 29: 636–64.

Earnhart, D. and Lizal, L. (2007) 'Does Better Environmental Performance Affect Revenues, Cost, or Both? Evidence From a Transition Economy', *William Davidson Institute Working Papers Series 856*, William Davidson Institute, MI: University of Michigan Stephen M. Ross Business School.

Frondel, M., Horbach, J. and Rennings, K. (2008) 'What triggers environmental management and innovation? Empirical evidence for Germany', *Ecological Economics,* 66: 153–60.

Hahn, R. and Stavins, R. (1994) 'Economic incentives for environmental protection', *American Economic Review*, 82: 464–68.

Horbach, J. (2008) 'Determinants of environmental innovations. New evidence from German panel data sources', *Research Policy*, 37: 163–73.

Huppes, G., de Koning, A., Suh, S., Heijungs, R., Oers, L., va Nielsen, P. and Guinee, J. (2005) 'Environmental impacts of consumption in the EU: High resolution input output tables with detailed environmental extensions', *Journal of Industrial Ecology*, 10: 129–46.

Jacobsen, H.K. (2000) 'Energy demand, structural change and trade: A decomposition analysis of the Danish manufacturing industry', *Economic Systems Research*, 12: 319–43.

Jaffe, A. and Palmer, K. (1997) 'Environmental regulation and innovation: A panel data study', *Review of Economics and Statistics*, 79: 610–19.

Jaffe, A., Peterson, S.R., Portney, P.R. and Stavins, R. (1995) 'Environmental regulation and the competitiveness of US manufacturing: What does the evidence tell us?', *Journal of Economic Literature,* 33: 32–63.

Johnstone, N., Hascic, I. and Popp, D. (2008) 'Renewable Energy Policies and Technological Innovation: Evidence Based on Patent Counts', *NBER Working Paper 13760*, Cambridge, MA: NBER.

Johnstone, N. and Labonne, J. (2009) 'Why do manufacturing facilities introduce environmental management systems? Improving and/or signalling performance?', *Ecological Economics*, 68: 719–30.

Kemp, R. (1997) *Environmental Policy and Technical Change,* Cheltenham: Edward Elgar.

Keuning, S., van Dalen, J. and de Haan M. (1999) 'The Netherlands' NAMEA; presentation, usage and future extensions', *Structural Change and Economic Dynamics,* 10: 15–37.

Martinez-Zarzoso, I., Bengochea-Morancho, A. and Morales-Lage, R. (2007) 'The impact of population on CO_2 emissions: Evidence from European countries', *Environmental and Resource Economics*, 38: 497–512.

Mazzanti, M. and Zoboli, R. (2006) 'Economic instruments and induced innovation: The European Directive on End of Life Vehicles', *Ecological Economics*, 58: 318–37.

Mazzanti, M. and Zoboli, R. (2009) 'Environmental efficiency and labour productivity: trade-off or joint dynamics?', *Ecological Economics*, 68: 1182–94.

Mazzanti, M., Montini, A. and Zoboli, R. (2007) 'Territorial productive structure and environmental efficiency indicators through regional NAMEA data', *Economia delle fonti di energia e dell'ambiente,* 49.

Mazzanti M., Montini, A. and Zoboli, R. (2008) 'Environmental Kuznets curve for air pollutant emissions in Italy: Evidence from environmental accounts NAMEA panel data', *Economic Systems Research*, 20: 279–305.

Milliman, S.R. and Prince, R. (1989) 'Firm incentives to promote technical change in pollution control', *Journal of Environmental Economics and Management*, 17: 247–65.

Popp, D. (2002) 'Induced innovation and energy prices', *American Economic Review*, 92: 160–80.

Popp, D. (2006) 'International innovation and diffusion of air pollution control technologies: The effects of NO_x and SO_2 regulation in the US, Japan, and Germany', *Journal of Environmental Economics and Management,* 51: 46–71.

Porter, M. (1991) 'America's green strategy', *Scientific American*, 264: 168.

Porter, M. and Esty, D. (1998) 'Industrial ecology and competitiveness: Strategic implications for the firm', *Journal of Industrial Ecology*, 2: 35–43.

Porter, M. and Van der Linde, C. (1995) 'Toward a new conception of the environment-competitiveness relationship', *Journal of Economic Perspectives*, 9: 97–118.

Reinhardt, F., Stavins, R, and Vietor, R. (2008) 'Corporate Social Responsibility Through an Economic Lens', *Resources for the Future working paper series no.12*, Washington DC: Resources for the Future.

Rennings, K., Ziegler, A., Ankele, K. and Hoffmann, E. (2006) 'The influence of different characteristics of the EU environmental management and auditing scheme on technical environmental innovations and economic performance', *Ecological Economics,* 57: 45–59.

Requate, T. (2005) 'Dynamic incentives by environmental policy instruments – a survey', *Ecological Economics*, 54: 175–95.

Requate, T. and Unhold, W. (2003) 'Environmental policy incentives to adopt advanced abatement technology: Will the true ranking please stand up?', *European Economic Review,* 47: 125–46.

Roca, J. and Serrano, M. (2007) 'Income growth and atmospheric pollution in Spain: an input-output approach', *Ecological Economics*, 63: 230–42.

Rubbelke, D. and Markandya, A. (2008) 'Impure Public Good Technologies and Environmental Policy', *FEEM Working Paper Series no. 76*, Milan: FEEM.

Schmidheiny, S. (1992) *Changing Course*, Cambridge, MA: MIT Press.

Stauvermann, P.J. (2007) 'The regionalized NAMEA-type matrix: Methodology, application and interpretation – A Dutch pilot study', mimeo.

Van den bergh, C.J.M. (2007) 'Evolutionary thinking in environmental economics', *Journal of Evolutionary Economics*, 17: 521–49.

Walley, N. and Whitehead, B. (1994) 'It's not easy being green', *Harvard Business Review*, May-June, 72: 46–52.

Ziegler, A., Schroder, M. and Rennings, K. (2008) 'The effect of environmental and social performance on the stock performance of European corporations', *Environmental & Resource Economics*, 37: 661–80.

Part I

Environmental innovation, firms' economic performances and policy drivers

1 Environmental innovation drivers and economic performance in industrial systems

Davide Antonioli, Massimiliano Mazzanti and Roberto Zoboli

Introduction

The issue of environmental innovation and its effects on overall firm performance and competitiveness in district-oriented local productive systems is particularly important in industrial areas with high density of firms, where clusters of firms may generate critical emissions and waste-hotspots. It is possible that these negative environmental features may be counterbalanced by the high propensity for innovation in district firms, where physical proximity enables them to exploit networking relationships and knowledge spillovers that dynamically increase the environmental efficiency of the productive area. Environmental innovative capacity is a current key issue. Environmental innovations are particularly crucial in local industrial frameworks since they often give rise to externalities with dual benefits: on the one hand, they provide the typical R&D (Research and Development) spillovers; on the other hand, they reduce environmental externalities (Rennings 2000; Jaffe *et al.* 1995). They contribute simultaneously to the Lisbon Objectives on growth and innovation and the Gothenburg priorities on sustainable development (IPTS 2004).

In relation specifically to manufacturing, the pollutants emitted by this sector are the main determinants of the pollution affecting the environment generally. Manufacturing industries account for the major part of total emissions. Evidence for Emilia Romagna, the northern Italian region considered in this chapter, suggests the importance of local industry concentration for local environmental pressure (Montini and Zoboli 2004). The high potential impact of this concentration depends on either the specific features of the production sector technologies or the spatial concentration of the industrial activity. Also, industrial districts (ID) (Brusco 1982; Brusco *et al.* 1996) are frequently in the top rankings for the most polluted local production systems.

Environmental innovation and its drivers

The empirical literature

In the literature on the drivers of environmental innovation,[1] Jaffe and Palmer (1997) made a seminal contribution. Their study provides an empirical investigation of the relationship between innovation and policy that is rooted in the Porter's (Porter and Van der Linde 1995) framework. Jaffe and Palmer find that higher lagged abatement costs lead to higher R&D expenditure.

Brunnermeier and Cohen (2003) employ panel data on manufacturing industries to provide evidence on the determinants of environmental innovation. They measure innovation by number of patents. They find that innovation responds to increases in abatement expenditures, but that the monitoring and enforcement activities associated with regulation do not have an impact in innovation strategies.

Evidence on environmental innovation in the European setting is provided by Frondel *et al.* (2004), who exploit OECD survey data for Germany in order to investigate whether environmental auditing schemes and innovations related to pollution abatement are correlated. Their main conclusions are that enhancement of the corporate image is a potential driver of the adoption of environmental management systems (EMS), while policy inputs do not seem to affect this organizational innovation. They find also that the public authorities' actions and the strictness of environmental policy seem to trigger abatement but that EMS and other policy instruments do not.

Rennings *et al.* (2006) also provide evidence for Germany, focusing on auditing schemes such as EMS and corresponding environmental organizational innovations. The main hypothesis they test is the influence of the maturity of EMS on environmental innovation processes. They find that EMS has a positive effect on all three forms of environmental innovation at firm level, with a key role played by the R&D department. Firms achieving significant learning success with EMS also show better economic performance. McKeiver and Gadanne (2005) analyse the external and internal factors that may affect the adoption of formal and informal audit schemes, such as EMS and ISO 14000, by small and medium enterprises (SME). Informal auditing conceptually links the realm of eco-auditing with the realm of environmental management and technology in the broader meaning. Biondi *et al.* (2000) propose an evaluation of the efficiency and effectiveness of EMS/ISO management tools. Such schemes, on the one hand may increase the complementarity of technological innovations (Mazzanti and Zoboli 2008), and on the other hand may increase the market value added for firms, by increasing the mark-up on the average costs that consumers are willing to pay.

This short review shows that incremental value is provided by analyses dealing with unexplored factors, such as networking, local stakeholder involvement, diverse policy proxies, environmental R&D, which play different and critical roles in SME environments. We frame and categorize innovation drivers as exogenous, endogenous (firm strategy), and the local idiosyncratic context in which the firm is embedded. A very particular case of embeddedness (Granovetter 1985), on which we focus in this chapter, is represented by a firm belonging to a local production system, for example, an industrial district (ID) (Brusco 1982; Brusco *et al.* 1996; Becattini 1990).

The analysis: data and context

We ground our applied analysis in a district-based local manufacturing system in Emilia Romagna, an area of Northern Italy characterized by a high density of ID, and a high level of per capita gross domestic product (GDP). Our dataset is very detailed. It is based on two surveys conducted on the same group of firms in 2002 and 2004, which elicited data concerning respectively 1998–2001 and 2001–2004. It is conceived as a quasi-panel and thus suffers from minimal cross-sectional bias, that is, the potential co-causal correlation between dependent and independent variables. It should be noted that evidence based on firm-level survey data of this richness and representativeness is rare (Khanna and Anton 2002).

The firms in the universe in this study (257 in 2001) belong to the manufacturing sector and have at least 50 employees. Reggio Emilia is one of the core SME-based industrial development areas in the region. The first survey carried out in 2002 consisted of a structured questionnaire that was administered directly to firm management and focused mainly on high-performance work practices (HPWP), industrial relations and technological/organizational innovations.

The 2004 survey on environmental issues was carried out by means of a short focused questionnaire administered to the 199 firms who had responded to the first survey (Getzner and Ritt 2006). Follow-up telephone interviews were conducted in November 2004. Participation in the second survey was 140 of the 199 firms, and showed no significant distortion by sector or size (see Table 1.1), with respect to the population. A proof of the good sample representativeness derives from the Cochran (1977) test.

The second questionnaire elicited information on process and product technological innovations introduced in 2001–2003, aimed at increasing environmental efficiency in emissions production, waste production and management of material inputs and energy sources.[2] We asked whether such innovations were produced within the firm, were based on co-operative agreements with other firms and/or co-operative agreements with research institutions, or alternatively were acquired from other firms. We also enquired about whether the innovation was associated with patenting activity and whether it resulted in the adoption of a corporate environmental management scheme. In terms of environmental policy, we asked about compliance with emissions and waste/energy policies and for how many years such policies had been in force. Other questions enquired about environmental R&D expenses, capital investments and policy costs (current costs, tax payments) for 2001–2003. Finally, firms were asked for details of any government environmental grants/subsidies to support eco-innovation.

A key value added of the *analysis* in this chapter is the focus on environmental performance and its drivers in SME. SME-based economic systems have certain features: (i) high density of medium-sized firms; (ii) large number of district firms,

Table 1.1 Interviewed firms and representativeness

	Number of employees						
Sector	*50–99*	*100–249*	*250–499*	*500–999*	*> 999*	*Total (%)*	*Total (absolute value)*
Food	0.00%	0.00%	1.43%	1.43%	0.71%	3.57	5
Other Industries	0.71%	0.00%	0.00%	0.00%	0.00%	0.71	1
Paper–publishing	2.14%	0.00%	2.14%	0.00%	0.00%	4.29	6
Chemical	3.57%	2.86%	0.00%	0.00%	0.71%	7.14	10
Wood	0.00%	0.00%	0.00%	0.00%	0.00%	0.00	0
Machineries	27.14%	17.14%	4.29%	2.86%	5.00%	56.43	79
Non-Metal Minerals							
(Ceramic)	10.00%	8.57%	2.86%	1.43%	0.71%	23.57	33
Textile	2.14%	1.43%	0.71%	0.00%	0.00%	4.29	6
Total (%)	45.71	30.00	11.43	5.71	7.14	100.00	
Total (absolute value)	64	42	16	8	10		140
To determine sample size: $n = N/[(N-1)\,\theta^2+1]$	sampling error						
Survey 2002 (n=199, N=257)	0.046						
Survey 2004 (n=140; N=199, 257)	0.04–0.055						

Note: n = sample size, N = universe/population, θ = sampling error.

characterized by a few, but strong production specializations; (iii) idiosyncratic industry climate with a pervasive role of unions in industrial relations, whose interactions with management over strategic decisions, can affect innovation dynamics (Mazzanti *et al.* 2006).

In addition to this focus on SME, we shed light on a diversified set of rarely investigated formal and informal drivers of environmental innovation, such as environmental R&D, policy-induced costs, role of industrial relations in innovation adoption, innovation-oriented networking activities, techno-organizational innovations and HPWP. We also exploit real performance indicators, based on the dataset of firm balance sheets registered with the Reggio Emilia Chamber of Commerce.

We refer the reader to Tables 1.2 and 1.3 for the descriptive statistics for the main environmental factors investigated.

Sector effects prevail over size effects on both the input and output sides of the innovation process. Environmentally critical sectors, such as chemicals, ceramics and paper, show more innovative dynamics. Medium and medium-large firms are the most involved, but based on type of innovation and index the picture is quite heterogeneous. However, it should be noted that a stronger structural break would have emerged if SME firms with less than 50 employees had been included, as these represent 50 per cent of the productive structure.

Note also that the share of firms reporting environmental-related patenting activity is very low (2 per cent). Although this is compatible with the historically

Table 1.2 Environmental innovations, R&D and environmental costs: descriptive statistics

Main Indicator variables	*Type*	*Mean*	*Max*	*Min*
Adoption of any environmental innovation	Dichotomous	0.79	1	0
Adoption of emission reduction related innovations	Dichotomous	0.49	1	0
Adoption of waste management related innovations	Dichotomous	0.42	1	0
Adoption of energy reduction related innovations	Dichotomous	0.46	1	0
Adoption of material input reduction related innovations	Dichotomous	0.27	1	0
Synthetic index of the adoption of environmental innovations	Between 0–1	0.41	1	0
Environmental R&D	% turnover, all firms*	0.55	10	0
Environmental Investments	% turnover, all firms*	0.78	10	0
Environmental costs	% turnover, all firms*	0.67	16	0
Environmental Patents	Dichotomous	0.02	1	0
Auditing Schemes (EMS or ISO)	Dichotomous	0.26	1	0

Note: *including all firms, with positive and zero values.

Table 1.3 Dimensional and sector effects: descriptive summary (mean values)

	Innovation (at least one form)	*Emissions reduction innovation*	*Waste related innovation*	*Energy related innovations*	*Material reduction innovations*	*Four innovations composite index (0–1)*	*R&D**	*R&D>0*	*Investments**	*Investments>0*	*Costs**	*Networking index (0–1)§*
<100 employees	71.88%	39.06%	34.38%	37.50%	25.00%	0.340	0.71%	37.50%	0.70%	42.19%	0.47%	0.152
100–249	80.95%	47.62%	50.00%	42.86%	21.43%	0.405	0.32%	45.24%	0.75%	61.90%	0.90%	0.208
250–499	93.75%	75.00%	50.00%	68.75%	31.25%	0.563	0.42%	56.25%	0.85%	68.75%	0.42%	0.172
500–999	87.50%	62.50%	37.50%	87.50%	50.00%	0.594	0.73%	37.50%	2.15%	62.50%	0.19%	0.281
> 999	80.00%	60.00%	50.00%	50.00%	50.00%	0.525	0.56%	60.00%	0.23%	30.00%	1.81%	0.225
Chemical	80.00%	60.00%	50.00%	50.00%	50.00%	0.525	1.30%	60.00%	1.36%	40.00%	0.58%	0.350
Machinery	79.75%	49.37%	35.44%	44.30%	29.11%	0.396	0.48%	40.51%	0.44%	49.37%	0.47%	0.184
Ceramic	81.82%	42.42%	57.58%	60.61%	21.21%	0.455	0.64%	60.61%	1.29%	66.67%	1.20%	0.182

Notes: * % of firm turnover, all firms included; the first five columns report the share of firms adopting such innovations. § networking index takes value one when firms cooperate for developing innovation for all four environmental realms considered; then takes values 0.75; 0.50; 0.25; 0.

small number of patents produced by Italian firms, there may be an incentive in district-oriented local systems characterized by a majority of SME not to patent innovation-related innovations given the uncertainties surrounding the defence of intellectual property rights. Other – perhaps informal – ways of protecting innovation rents may be more common, perhaps related to idiosyncratic, non-transferable features of the organizations involved. Thus, unlike other studies on the determinants of innovation (Brunnermeier and Cohen 2003), in this study patenting is not the best proxy for innovative capacity and adoption. Firms may choose to protect their property rights by means of trade secrets and copyright rather than patents (Gu and Tang 2004).

Examining the factors behind environmental innovations

The analysis in this chapter provides new evidence on the driving forces of environmental innovation by investigating the adoption of product/process innovation as outputs and R&D as an input and testing a set of hypotheses. Conceptually, we sub-divide the set of relationships defining exogenous, idiosyncratic and endogenous (strategic management related) factors. Our primary focus and comments relate to the main added value lines on which the analysis is grounded, that is, networking activities, industrial relations (the main idiosyncratic factors), and policy-related effects. This approach is linked to the rise in heterodox approaches that claim that socio-political considerations and internal/external organizational features influence firms' innovative dynamics over and above the mere effect of environmental economic instruments often assumed in mainstream perspectives (Kemp 2000). Co-evolutionary approaches and the behavioural theory of the firm also provide useful and sometimes better views of the theory of innovation drivers (Krozer and Nentjes 2006). The effect of networking, which we define as having hybrid features (external organizational change and idiosyncratic element), is discussed in relation to R&D. It should be noted that some of the factors investigated also may present hybrid features (e.g. idiosyncratic, but related to the firm strategy, such as networking).

Table 1.4 presents the set of hypotheses investigated. For an extensive discussion of these hypotheses regarding output and input innovation drivers, and

Table 1.4 Summary of main hypotheses

	Hypothesised influencing factors	*Incidence on environmental output innovations*
H.1	Group membership	(+, –)
H.2	Industrial relations	(+, –)
H.3–4	Organisational/technological innovations, EMS	(+, –)
H.5	Policy actions / Policy induced costs/grants	+
H.6	R&D	+
H.7	(Past) firm performances	+

an econometric exercise involving R&D drivers, we refer to Mazzanti and Zoboli (2009).

There is no dominant model for studying the determinants of innovation at either industry or firm level (Jaffe and Palmer 1997). At a conceptual level, in this chapter we extend the usual linear innovative process, which tends to link innovation to R&D as an input, to a richer, more developed innovation production function. We would suggest that when studying output innovation and input proxies from an applied perspective, it is plausible to define reduced forms that try to explain innovation by exploiting a broad but theoretically consistent set of covariates.

We can claim that this study is robust based on its sample representativeness and the quality and quantity of the firm level data. The rich set of information mitigates the possibility of endogeneity that can derive from omitted variables (correlation between explanatory variables and errors) due to data unavailability. Endogeneity can arise when investigating the determinants of innovation and panel datasets may provide a better framework in this case. One way to deal with the problem is to introduce into the regression a vector of the lagged term, for all or some covariates (Khanna and Anton 2002). In exploiting two survey waves, most of our covariates temporally precede the innovations identified (2001–2003). We test for endogeneity for key variables, such as R&D and costs, using the Wu–Hausman test (Wooldridge 2002). We discuss some other econometric issues and robustness tests below.

First, we conduct a preliminary analysis to study the full correlation matrix. This first selection is aimed at reducing collinearity problems. The outcome is a matrix of selected potential explanatory variables.

We apply a general to particular backward stepwise method: over-fitting specifications starting from a conceptual model is a less flawed method and can be resolved by (potential) step-by-step deletion of non-significant variables.

Second, in estimating the total innovation index, ranging between zero and one, we face a limited, but continuous variable. We deal with continuous, but limited fractional variables. It is possible to verify that estimates deriving from ordinary least square (OLS), OLS (log) transformations and Tobit models do not differ significantly (Pindyck and Rubinfeld 1991). Since the aim (here) is not the estimation of elasticity values, we use OLS corrected for heteroskedasticity as an econometric tool.

Econometric analysis focuses on the determinants of: (a) a specific environmental output innovation (emission-related, waste-related, and energy related); (b) a synthetic index of all adopted output innovations.

We estimate a sort of knowledge production function (Griliches 1979). A knowledge production function expresses the relationship between output innovation and innovation inputs. The reduced forms for the two levels of analysis are as follows:

$$\begin{aligned} INN_{i,t} = \; & \beta_0 + \beta_{1,t}(\text{structural firm features}) \\ & + \beta_{2,t}(\text{environmental policy proxies}) \\ & + \beta_{3,t}(\text{environmental R\&D}) \\ & + \beta_{4,t-1}(\text{techno-organizational innovation}) \\ & + \beta_{5,t-1}(\text{industrial relations}) + \beta_{6,t-1}(\text{past performances}) + e_i \end{aligned} \tag{1.1}$$

where INN_i represents the environmental output innovation index of firm i[3] and e_i is the error term with the usual properties. β_0 is the constant term and β_{1-6} is the set of coefficients associated with the explanatory variables, where (t) is 2003–2001 and (t–1) is 2001–1998.

Econometric results

We present and comment on the results for the set of hypotheses formulated in Table 1.4. We examine various environmental-related output innovation equations.

We test for potential endogeneity, although as noted above: (i) the emphasis is on trends – which is plausible given the slow-evolving nature of the variables; (ii) the causality nexus in this case is clearer than innovation–performance links in pure cross-section frameworks (Michie and Sheehan 2005). In fact, R&D and costs are conceptually inputs, and auditing schemes may be correlated to, but hardly explained by technological innovations. Tests show that exogeneity is rejected only in the case of costs, but not R&D and EMS. This confirms ex-ante expectations, since, relatively speaking, induced costs were the factor most likely to present endogeneity problems. In such cases we introduce the associated fitted values as a further two-stage estimation case (Wooldridge 2002). We introduce R&D and costs as explanatory variables both separately and jointly, to check whether their positive correlation may lead to distortions in the estimates.

The outcomes of our output innovation proxies are presented in Table 1.5. First, environmental innovations related to emissions reductions are positively influenced by the presence of voluntary auditing schemes. In terms of policy-related explanatory factors, we note that the (stated) presence of an emissions policy is positively related to innovation; nevertheless, and interestingly, the probability of producing an emissions-related innovation is inversely proportional to the number of years that a firm has been subject to an emissions policy. This number of years, reported by the firms themselves, may be related to historical, productive or institutional reasons. The outcome may seem counter-intuitive and is confirmed below. Based on the evidence it seems that policy effects are stronger in the first phase of its implementation and reduce over time. However, it might be that because we are observing 2001–2003 innovations, most firms have already produced innovations based on policy, and the innovative firms we identify may be newcomers.

The index of participative innovation-oriented industrial relations is one of the positive drivers. The positive effect of R&D arises only if we specify a dummy

Table 1.5 Econometric regressions (innovation adoptions)

Dependant variable	*INNO–EM*	*INNO–WA*	*INNO–EN*	*INNO–TOT*	*INNO–TOT*	*INNO–TOT*
Regression	1	2	3	4	5	6
Covariates / Methodology	Probit corrected for heteroskedasticity	Probit corrected for heteroskedasticity	Probit corrected for heteroskedasticity	OLS corrected for heteroskedasticity	OLS corrected for heteroskedasticity	OLS corrected for heteroskedasticity
Constant	–0.945	–1.392	–2.676***	0.941	0.135	0.083
Log-Size	–0.229	–0.754	1.514	0.416	0.196	0.272
CHEM	0.456	0.605	1.846*	1.668*	1.778*	1.579
MACH	–0.149	0.256	1.645*	0.619	0.720	0.547
CERAM	–1.678*	1.822*	2.234**	1.186	1.223	1.318
GROUP		1.971**		1.515	1.758*	1.982**
HYER		–2.078**	–1.125	–1.892*	–1.831*	–1.786*
IND_REL	2.397**		2.546**	2.477**	2.492**	2.293**
POL-WA/EM	2.090**	2.857***				
POL- WA/EM (YRS)	–2.243**	–2.304**				
GRANT		1.916*		3.707***	3.194***	3.670***
ENV-INV			(dummy) 2.115**	–0.975		
ENV-COST		1.752*		2.794***	2.397**	
ENV-COST (pred values)	Not significant when included	Not significant when included			Not significant when included	
R&D				2.131**		2.535**

R&D dummy	2.081**		Significant at * when included			
AUDIT	2.185**	2.768***		3.076***	2.951***	3.038***
EMS				EMS significant at *** when included separately		
ISO 14000						
$PROD_{9800}$		1.302				
McFadden pseudo R^2	0.158	0.216	0.154			
Estrella fit	0.213	0.282	0.206			
Adj R^2				0.192	0.200	0.194
Log-L	–81.56	–81.75	–81.75			
χ^2 LR test (prob χ^2>value)	0.0006	0.00004	0.0002			
F test (prob)				3.21 (0.0002)	4.17 (0.0000)	4.05 (0.0000)
Correct prediction: actual 1s and 0s correctly predicted	70%	75%	67%			
N	140	140	140	140	140	140

Notes: The table presents t ratios since we are not interested in coefficients that do not represent elasticity in such analysis. Empty cells represent the variable is not included in a given specification following the "from general to particular" procedure we discuss in the text. *** p values < 0.01, ** p values < 0.05, * p values < 0.1.

variable as an explanatory factor (neither R&D/employees nor environmental costs are significant). Size and sector had no influence on adoption, and firm performance also does not matter.

Second, waste management related innovations are affected primarily by policy proxies. As before, we note that while the policy dummy is positively significant, the probability of adopting a waste management innovation is inversely related to the number of years since the policy was implemented. Although the number of firms exploiting grants is low, it is significant in this case. Waste management innovations are shown to be positively influenced by the presence of voluntary auditing schemes and by a flatter organizational structure. Group membership overturns size effects, and some sector influence emerges.

Third, in the realm of energy efficiency, we observe that R&D, included as a dummy variable, is a significant endogenous firm driver. In this case, capital investments are significant for explaining energy innovations: this is plausible given the high technology fixed costs and the low relevance of end-of-pipe solutions. Again, the effect of industrial relations dynamics is positive. Also, sector has an influence on innovation: this is not really surprising given the differences in energy intensity across manufacturing sub-sectors.

Finally, we examine the index capturing all four realms of innovation. We note (Table 1.5, last 3 columns) that: (i) R&D and costs are significant while investments are not (regression 4); (ii) policy drivers, such as grants, and policy-driven induced costs are significant. Induced costs are not significant when included as predicted values. This undermines the evidence in favour of an effect of induced costs on innovation. Auditing schemes are significant (with EMS effects more important than ISO 14000). Sector and size do not influence the adoption of innovation. Scale economies emerge through the effects of group membership. Finally, and confirming our earlier evidence, innovative activity is more intense in flatter organizations and firms where industrial relations are good in terms of workers' and union participation in decision making about high-performance and organizational strategies. Past performance is shown not to influence environmental innovation.

Complementarities, firm strategy and innovation

The literature

The relevance of complementarity among the drivers of performances has been underlined by works in the wide-ranging literature on the relationship between innovation strategy and firm-level performance. Since the mid-1990s, contributions have highlighted the limited, short-run effects of strategies biased towards organizational (cost-) efficiency and the higher potential for increasing long-run performance through innovation-based management (Huselid 1995; Black and Lynch 2001; Ichniowski *et al.* 1997). We refer to Mazzanti and Zoboli (2008) and Mancinelli and Mazzanti (2009) for in-depth discussions of the methodological issues and related literature.

Analysis of complementarity includes diverse factors affecting firm performance such as technological innovation, R&D, organizational innovation, HPWP, training, networking. Many studies include empirical analyses, which explore various hypotheses with respect to the effects of complementarity on firm performance (productivity, profits) and firms' innovation performance. Complementarity is also studied from another, conceptual and empirical perspective, by more evolutionary, systemic-oriented and dynamic-focused streams of research, which offer some different and interesting insights (Teece 1996), for example, the idea of co-specialization between productive factors. In contrast to spillovers, if there is complementarity between two firm activities, this implies that were one of these activities to increase, it would be beneficial for the firm to increase the other activity: this produces a system effect in which the whole is more than the sum of the parts. This has obvious implications for both firm strategies and policy decisions.

In the presence of two factors[5] on which we aim to test for a complementarity link, this holds if and only if (11) – (10) ≥ (01) – (00),[6] that is, if the incremental value of a strategy that moves from one factor to two factors is higher than the value of a strategy that adds one driver starting from the state (00), or in other words, if complementarity characterizes those specific factors.

Hypotheses, methodology and analysis

To test for complementarity we test our hypotheses at two levels of reasoning. First [*Hc1*], we assess complementarity at the level of innovation adoption, analysing the correlation between adoption of environmental innovation in emissions, energy, waste, and materials. We expect that more innovative firms (the leaders in the local system) jointly adopt eco-innovations associated to different realms.

Secondly [*Hc2*], we investigate complementarity at the level of innovation inputs. Building on the above discussions around innovation drivers, we consequentially test the complementarity between voluntary auditing schemes with policy-driven costs and the firm's R&D strategy. We are interested in assessing whether such EMS/ISO are correlated to innovative R&D strategies and are co-evolving with policy-driven effects. This is crucial for informing environmental policy related to both sides: incentivizing R&D given potential market failures; supporting auditing for promoting environmental corporate effort and mitigating free-riding behaviour in regional contexts; imposing additional costs through a diverse set of policies – typically taxes on emissions, energy use and waste disposal. Further, we test the complementarity between internal strategies (R&D) and externally driven factors (costs) in stimulating innovations.

There are three possible methodologies. The first analyses complementarity by studying the correlation between two or more variables, controlling for other factors. This is usually accomplished by establishing a bivariate or multivariate probit model. In this case the variables under scrutiny are the dependent elements of the empirical model (Galia and Legros 2004; Laursen and Mahnke 2001).

The second is defined as a reduced-form approach (Arora 1996): the analysis of complementarity is carried out by focusing on the effects of two factors, and their correlation. It is typically implemented by setting interaction terms (Athey and Stern 1998).

The third approach allows the greatest flexibility and is the most popular. We describe it as the productivity approach: it can cope with two or more factors on which the hypothesis is tested, and is based on the estimation of an objective function, either a production or an innovation function (Laursen and Foss 2003; Brynjolfsson *et al.* 2002). The most common method used is to assess the hypothesis by testing the significance of interaction variables, which capture the complementarity effect. Alternatively, it is possible to analyse complementarity within a discrete framework. Going directly to the definition of complementarity in this discrete setting which analyses the hypotheses by pairs of factors, we note that complementarity holds if $(11) + (00) \geq (01) + (10)$, where (11) is the state of the world in the presence of both factors (innovation drivers), and so on. Theoretically, inequality is tested on a non-strict basis (meaning ≥ instead of >).[7]

A bivariate probit analysis is carried out to test the correlation between various environmental innovations (Table 1.6 presents the correlation values for the regressions). The adoption of emissions-reduction technology is correlated to the adoption of both waste- and energy-oriented technologies. Emissions reduction arises plausibly, therefore, as a leading strategy, coherently and expectedly linked to energy-saving technologies, but also to waste-oriented strategies. The reason for this may be that higher waste-management efficiencies strengthen or add to emissions strategies. It is possible that the more materials flows are reused and re-integrated within the firm's productive processes, the less emitting will be the overall process. Or, it could be that emissions and waste management/disposal, at the Italian and other levels, are two relevant policy targets. The observed correlation could be a response to greater policy emphasis in those areas. The lack of correlation between emissions and materials strategies may be a sign that end-of-pipe innovations, which predominate for emissions, and at source-oriented strategies, relevant to greater materials efficiency, are not integrated within the firm's environmental strategic planning.

Table 1.6 Bivariate probit analyses on innovation adoptions

Dependent variables	*Correlation coeff.*	*(T value)*
INNO-EM/INNO-WA	0.459	(3.720)***
INNO-EM/INNO-EN	0.58	(5.271)***
INNO-EM/INNO-MA	0.08	(0.574)
INNO-WA/INNO-EN	0.133	(0.947)
INNO-WA/INNO-MA	0.399	(2.898)***
INNO-EN/INNO-MA	0.274	(1.870)*

Notes: N=140; only firm structural characteristics and performances are used as covariates. Regression estimates are available upon request.

Waste processes are also correlated to materials input-reduction strategies; this confirms the clear expectation that waste-management strategies complement reductions in material flows in processes. Finally, we note that waste and energy are not correlated to innovation adoption, which is probably linked once again to the absence of an integrated process, which sets up waste-related energy saving investments, such as bio-mass/waste recovery by incineration, at the SME level.

Overall, the set of correlations confirms that innovation dynamics, both technological and techno-organizational, are generally highly correlated to each other. When firms become environmental innovation adopters they tend to pursue an integrated strategy covering diverse fields within the green arena.

For the objective function, we estimate regressions for specific dichotomous innovation proxies (energy, emissions, waste) and for the index of innovation intensity.

To assess complementarities, we specify regressions entering the four dummies associated with the states of the world for each state of presence/absence of the two factors: [00, 01, 10, 11]. All coefficients related to state dummies are estimated, thus specifications do not include a constant term. The null hypothesis is of complementarity under a non-strict inequality (≥0).

Estimates and consequential tests show that the complementarity hypothesis holds in most cases (Table 1.7). We note, nevertheless, that ENV-COST is associated with negative signs, but in only one case (R&D versus ENV-COST for the emission-related innovation specification) is there rejection of the null. R&D and policy costs, therefore, seem to be substitute drivers in this specific context, pointing to the use of one 'driver' only.[8] We also observe that the only case where complementarity would hold, even in a strict sense (> 0), is for AUDIT versus R&D, in the adoption of energy-saving innovations. AUDIT and R&D relationship signals an interesting complementarity, often debated at a conceptual level, between organizational and technological innovation.

Overall, we observe that complementarity, as defined above, holds for most specifications.

To sum up, the analysis confirms first, the need for careful verification of the hypothesis of complementarity for each contingent case under scrutiny, and second, that in our framework the main drivers of environmental innovation, R&D, policy-related costs and auditing envisaged comprise a framework in which complementarities are relevant, but trade-offs may emerge and may be plausible.

Table 1.7 Discrete based complementarity tests on innovation drivers

	Auditing / R&D	*Auditing / ENV-COST*	*R&D / ENV-COST*	*R&D / Networking*
INNOTOT	0.18	–0.63	–0.18	–1.93/2.33
INNOEM	0.19	–1.35	–2.00	–0.33/0.52
INNOWAS	–0.58	–0.84	1.05	–2.83/2.80
INNOEN	2.14	0.49	–0.24	0.10/1.10

Note: values of the T test (one-sided t test) shown.

Firms may exploit external resources to complement internal R&D efforts and achieve economies of scale in innovation strategies where resource interdependencies are required to implement radical change. The message based on our evidence is unclear. When we consider the innovation index, evidence emerges of strict complementarity if policy-related costs and auditing are also included. If only the controls are included, we find evidence of substitution. Thus, the outcome of our investigation is not robust to a sensitivity analysis on the vector of independent variables. As far as emissions and energy-saving innovations are concerned, our hypothesis of complementarity holds. Firms may achieve competitive advantage by integrating R&D, and networking with other firms and institutions in the case of SME, or by exploiting only internal resources (R&D) in the case of larger firms, or by exploiting external sources of innovation in the case of small firms. With regard to waste, the outcome is the same as for the synthetic index of innovation. This ambiguous evidence may in part be due to the non-radical nature of most of eco-innovation changes experienced by the innovation system so far.

Innovations and economic performance

A survey

This section provides a review of the literature. We focus first on the more general framework of techno-organizational innovations and their performance effects, and the role of industrial relations and other factors that may play a role in district-based local production systems, and then examine the specific issue of environmental innovations. The literature review highlights the state of the art and the main value added of our analysis: the role of SME in local systems, the role of local idiosyncratic factors, and an assessment of the effects of (eco-)innovations and other competitive strategies on economic performance.

It has been widely recognized that innovation activity is fundamental to a firm's sustained competitive advantage. It has been shown that innovation is a broad concept that encompasses, at least, technological and organizational aspects (OECD 2005).

The importance of maintaining a broad perspective on innovation is underlined by the now well-rooted concept of the knowledge-based economy. The diffusion of knowledge-intensive production technologies and organizational practices has shifted the attention of many economists to the importance of how knowledge is managed within firms (Kremp and Mairesse 2004; Hall and Mairesse 2006).

The conceptual background that justifies the adoption of our integrated view of innovation activities stems from works that can be categorized as of the knowledge-based perspective of the firm. In terms of the (dynamic) organizational capabilities and co-evolving assets and activities related to technological innovations (Teece 1996; Coriat and Dosi 2002), the importance is underlined of 'new organizational practices' coupled with and complementary to technological innovation.

Thus, new organizational practices (EC 2002), which require/imply a better skilled workforce, a flatter hierarchical structure, delegation of responsibility, and degree of decisional decentralization and autonomy in managing job tasks, coupled with technological innovations, contribute to sustaining firms competitive advantage (Caroli and Van Reenen 2001; Black and Lynch 2001; Lynch 2007).

Employees' human capital is a fundamental resource since organizational innovation requires a strong skills base (Leiponen 2005). This highlights the importance of training activities (Zwick 2004; Conti 2005), which help the generation and accumulation of the skills and competences required to technological innovation. In other words, the firm's organizational design should include human resource management practices that imply higher level of commitment and involvement from employees.

A related element that is idiosyncratic and rather under-explored and which is linked to organizational and human capital strategies and has the potential to influence firm performance, is industrial relations. Trade unions can enhance or hinder workplace productivity and profitability through their influence on innovation activities (Booth 1995; Menezes-Filho and Van Reenen 2003; Metcalf 2003; Menezes-Filho *et al.* 1998; Machin and Wadhwani 1991).

Although the insights in the conceptual literature are not unambiguous (Menezes-Filho and Van Reenen 2003; Metcalf 2003), we generally find evidence of a neutral or positive influence of unions on productivity and a negative influence on profitability. Whether the positive effect outweighs the negative effect, thus qualifying unions as surplus-enhancing, needs to be analysed case by case. In fact, if industrial relations are cooperative, we should expect a more committed workforce, higher morale, and a more stable environment, all with positive consequences for firm efficiency. If, on the other hand, industrial relations are adversarial, and the actions of unions conflict with the company's interests, management may ignore them resulting in no alignment between union and management goals, which we would expect to have negative effects.

The literature on the drivers of eco-innovation and related effects on performances, is inspired on the one hand by Porter's hypothesis and framework of environmental regulations effects on firm performance and innovation paths, and on the other hand by the literature on corporate social responsibility (CSR). CSR is seen as a strategic behaviour (Portney 2008; Lyon and Maxwell 2008), which, via investment in innovation, in the long run may bring together the private and public elements of benefits (e.g. profits, public goods). Debate revolves around the possibility of integrating firm performance and the provision of (environmental) public goods, using the lever of (eco-)innovation activities to try to mitigate the trade-offs or transform trade-offs into complementary dynamics.

Eco-innovation is becoming the conceptual reference point for many regional and international public policies and management strategies and has been defined as the production, application or use of a product, service, production process or management system new to the firm adopting or developing it, and which implies a reduction in environmental impact and resource use (including energy) throughout its life-cycle. This includes innovations whose environmental effects are unintended.

Studies of environmental innovation and its relationship with socio-economic performance originate within three analytical frameworks of environmental economics and management/innovation studies.

Following Porter's work, many authors have addressed the effects of competitiveness on manufacturing sectors most vulnerable to competition and regulations (Jaffe *et al.* 1995), and the motivations for CSR – whether they are profit-only or aimed at the production of a partially public good (Reinhardt *et al.* 2008). It is crucial to understand: whether the firm is being merely compliant with existing regulations or is going beyond them; whether it would have pursued innovation even in the absence of regulations; how costs and benefits are distributed along the dynamic path (from short to long term) that follows innovation.

For a detailed critique of Porter's hypothesis see, among others, Jaffe *et al.* (1995), which empirically and theoretically addresses the hypothesis that environmental regulation may benefit the affected firms.

The analysis: data and research hypotheses

In order to test a varied set of research hypotheses, we exploit data deriving from the two surveys described earlier and data on official balance sheets for the year 2004, which provides figures for productivity in terms of value added per capita. The availability of balance sheet data for SME is a major value added of the empirical analysis in this chapter. Our dataset of 144 firms derived from the intersection of the first two surveys is merged with the balance sheet data, leading to a final sample of 89 firms with unique characteristics in the field of innovation and environmental studies.

Table 1.8 presents the signs of the relationships among the sets of covariates and the productivity variable.

First, we expect a positive effect on labour productivity of techno-organizational innovation and HPWP generally including training. Second, the relevance of management–union and management–worker relationships in highly unionized industrial areas, such as the districts in Emilia Romagna, cannot be ignored. The assumption is that good (cooperative) industrial relations and good (participative)

Table 1.8 Hypothesized signs of influencing variables on the productivity (VA/E)

Influencing factors	*VA/E*
Techno-organizational innovation / HPWP (Training, process/product technological innovation, high involvement work practices, plant flexibility, etc.)	+
(Cooperative) industrial relations	+
Environmental innovations (ISO; EMS, R&D, adoption of process/product innovations)	+/–
Environmental costs/environmental policy	+/–

worker involvement in some business strategies entailing techno-organizational innovations and HPWP, could induce higher performance in such firms.

The role of environmental-related activities on labour productivity is more ambiguous and consequently we expect both positive and negative signs. On the one hand, if we focus on the role of eco-innovation factors in stimulating economic performances, from a Porter/CSR perspective (Reinhardt *et al.* 2008), we would hypothesize that eco-innovation strategies produce positive effects on economic performance. On the other hand, the adoption of labour productivity could be driven merely by non-strategic reasoning (compliance with policy), and divert resources from other productive investments.

The last factor we investigate is related to the impact of policies on firm performance. We test this link, which may be positive, not significant or negative depending on the same reasoning described above. In fact, environmental regulations can spur positive private benefits for firms in the medium-long run, only if these firms react by innovating and being proactively socially responsible. Because the time lag we analyse is embedded in the short run, we would expect negative or negligible effects on productivity.

In order to test the above hypotheses we construct a hybrid cross-section regression model that exploits the lag between economic performance and innovation drivers.

As a general specification of the econometric model we estimate the following form:

$$\begin{aligned}\text{Prod (VA/E)}_{i,04} = \; & \beta_0 + \beta_{1i,98\text{–}01}\,[\text{structural variables}] \\ & + \beta_{2i,98\text{–}01}\,[\text{techno-organizational innovations/ HPWP}] \\ & + \beta_{3i,98\text{–}01}\,[\text{industrial relations}] \\ & + \beta_{4i,01\text{–}03}\,[\text{environmental innovations}] \\ & + \beta_{5i,01\text{–}03}\,[\text{environmental policy/cost}] + \varepsilon_i \end{aligned} \tag{1.2}$$

where β is a vector of the coefficients for each set of covariates, the dependent variable refers to 2004, i identifies each firm, the sets of covariates refer to both 2001–2003 and 1998–2001.

The variables in the above reduced form can be summarized as follows. The productivity indicator is constructed on the basis of balance sheet information for 2004. The ratio of firm value added to number of employees (VA/E) is an expression of labour productivity. For detailed information on the construction of the variables, see Table 1.9,[9] and Antonioli and Mazzanti (2009).

Drivers of productivity: innovation strategies and environmental factors

First, from the set of innovations and HPWP it emerges that structural features, such as size and sector, are not significant (Table 1.10, column 1), and thus are

Table 1.9 Descriptive statistics and variable contents (89 firms)

Variables Acronym	*Variables Description*	*Min*	*Max*	*Mean*
	Dependent			
VA/E	Logarithm of value added over employees for the year 2004	7.91	173.73	54.91
	Structural Variables			
Sectors Dummies (Chemical, Machineries, Non-metallic mineral products)	Binary variables (0,1)	0	1	/
SIZE	Number of employees in 2004	50	2239	233.50
GROUP (d)*	Belonging to a group: binary variables (0,1)	0	1	0.30
NAT_TURN	Percentage of turnover made on national markets	0	1	0.48
MKTREV	Percentage of production sold on the market	0	1	0.83
SK_RATIO	Skill ratio: managers, middle-managers and specialized blue collars over clerks and not-specialized blue collars	0	3.36	0.85
	Techno-organizational Variables			
INNO_PROD (d)*	Product innovation: binary variable (0,1)	0	1	0.03
INNO_PROC (d)*	Process innovation: binary variable (0,1)	0	1	0.71
OUTS	Index that captures the diffusion of outsourcing strategies for 7 firm ancillary activities. Interval (0–1) $\sum$ Outsourced Activities / $\sum$ Total Activities	0	1	0.28
PLANT_FLEX	Plant flexibility. Presence of flexible polyvalent plants both in presence of flexible/adaptable labour activities and rigid/predetermined labour activities. Interval (0–1)	0	1	0.36
INNO_ORG	Index that captures the diffusion of team working, TQM, JIT, quality circles and job rotation. Interval (0–1) $\sum$ Implemented Practices / $\sum$ Total Practices	0	1	0.25
HIERARC_INT	Index of hierarchical intensity. Interval (0–1) $\sum$ Numb. Hierarchical Levels / $\sum$ Numb. Formalized Function	0.13	1	0.28

COV	Index showing the percentage of employees involved in training programmes (0 nobody; 1=1–24%; 2=25–49%; 3=50–74%; 4=75–100%). Rescaled on (0–1) interval	0	1	0.48
BONUS	Index that captures the presence and diffusion of bonus schemes, according to the different employees categories: managers, middle-managers, clerks, specialized blue collars and not-specialized blue collars. Interval (0–1)	0	1	0.33
INNOPART	Index capturing the diffusion of consultation and delegation processes among individual employees or groups. Interval (0–1)	0	0.70	0.25
	Industrial Relations Variable			
IND_REL	Industrial relations index that informs on the intensity in cooperative behaviour between management and unions at firm level. Interval (0–1)	0	0.87	0.33
	Environmental variables			
EMAS (d)*	Environmental innovation to reduce emissions: binary variable (0,1)	0	1	0.03
ISO 9000(d)*	Formal certification ISO 9000: binary variable (0,1)	0	1	0.19
ISO 14000(d)*	Formal certification ISO 14000: binary variable (0,1)	0	1	0.15
RSN	Environmental R&D expenditure per employee (€M)	0	0.05	0.001
CERTIF(d)*	Auditing schemes: binary variable (0,1)	0	1	0.30
ENVCOST	Current expenses in environmental management and policy compliance (e.g. taxes, permits)	0	0.02	0.001
YRSWAS	Number of years the firms has been subject to waste policies	0	25	7.57
YRSEMIS	Number of years the firms has been subject to emission policies	0	26	8.92

Note: (d) means dummy variable

Table 1.10 Labour productivity effects of (eco-) innovation strategies

Dep.var. value added/employees	*1*	*2*	*3*	*4*	*5*
Constant	24.31**	19.59*	23.65**	23.36**	23.40**
Structural variables					
Firm Size	Yes	Yes	Yes	Yes	Yes
Sector dummies	Yes	Yes	Yes	Yes	Yes
GROUP	–14.59***	–16.54***	–12.21***	–12.18***	12.08***
MKTREV	17.98***	17.19***	16.65***	16.64***	12.88**
Techno-organizational variables/HPWP					
COV	13.58*	16.18**	12.79**	13.02**	11.95**
INNORG	21.52**	/	15.82*	14.64*	21.47**
INNOPROC	–7.51*	–8.26**	–7.81*	–7.45*	/
REWARD	28.30***	27.05***	26.24***	26.54***	22.76***
INNOPART	/	41.76**	/	/	/
Environmental variables					
EMAS	/	/	7.69	/	/
ISO 14000	/	/	16.57*	17.84**	18.92**
RSN	/	/	/	/	586.15***
WASPOL	/	/	/	/	–0.864***
N	89	89	89	89	89
F test	3.43 (0.0006)	3.62 (0.0004)	3.86 (0.0001)	4.19 (0.0000)	5.00 (0.0000)
Adj R^2	0.233	0.247	0.296	0.302	0.371

Notes: OLS corrected for heteroskedasticity is used; *** p values < 0.01, ** p-values < 0.05, * p-values < 0.1; SIZE and sector dummies (Chemical, Machineries, Non metallic mineral products) are included but not shown as not significant. / means not included, following mainly a pre-analysis on correlations among covariates, with the aim of mitigating collinearity.

omitted from the table. Only the coefficient for share of market-related revenue (as opposed to mainly subcontracting) shows a positive significant sign. It is the firm's strategy in relation to product markets, not size or manufacturing sector, that matters.

It is interesting and somewhat counter-intuitive that group membership is related to productivity through a significant negative link: the reason may be that the association between size and economic performance in local systems is often bell shaped. Medium-sized firms perform better; large firms are mainly involved in group strategies, making our evidence consistent with other evidence on this manufacturing system (Antonioli *et al.* in press).

Within the bundle of innovative and HPWP strategies, training and 'organizational innovative actions' are more significant. Specifically, training coverage (COV), the additive total index of HPWP (INNO_ORG), embedding quality circles, total quality management (TQM), team working, just-in-time, job rotation), adoptions and the use of monetary bonuses for employees' efforts (BONUS) emerge as the main drivers.

Counter-intuitively, adoption of product innovations does not impact on productivity, and process innovation is associated with a negative and moderately significant coefficient. However, this is consistent with recent investigations into the same local production system (Antonioli *et al.* in press): technological innovations as performance drivers are ranked third, lower than training and organizational innovation. In line with expectations, training and organizational innovation have an impact on labour productivity. Overall, organizational and human capital oriented strategies seem to matter for economic performance. The lack of impact of technological innovations may be due to the time lag before their impact on productivity. The negative sign of INNO_PROC could be related to the diversion of investment for the adoption of the innovations, which, in the short run, crowd out other productive value-creation strategies that are more dependent on current technological assets. Finally, in line with other similar studies, we analyse adoption in terms of the binary variables. The intensity and 'radicalness' of innovation are aspects that future research could investigate.

Second, in analysing the single HPWP elements included in INNO_ORG, we observe that its significance is driven by team working, just-in-time, and quality circles, which are all highly significant, but not by TQM or job rotation (for reasons of space we do not show the ancillary regressions with separated effects). The reason why TQM is not significant may be due to the quite general content of this HPWP, which may be formally rather than effectively adopted. For the index capturing intensity of employee involvement (INNO_PART) we find it is positively linked to labour productivity (Table 1.10, column 2), which is in line with the findings from other studies (Addison *et al.* 2000; Pencavel 2003). Shifting some decisional power to employees and endowing them with greater autonomy and responsibility through an organizational re-design seem to encourage higher labour productivity. In terms of the results for INNO_PART, the index of 'good' industrial relations (IND_REL) is not significant, and is not reported here. This result is not completely unexpected: other studies of Reggio Emilia find similar effects of industrial relations on firm performance, in terms of its impact on innovation intensity at firm level (Antonioli 2009).[10] Good industrial relations is a factor, which (by acting as a stimulus to innovation) may indirectly rather than directly increase firm performance. In the first specification we note that in column 2 of Table 1.10 both the F test and the R^2 increase in value after the inclusion of more covariates, the constant loses its explanatory power, and training (COV) shows a bigger impact; the other evidence remains the same. INNO_ORG is omitted in this case because of its high correlation with INNO_PART.

Third, we examine the role of environmental strategies (Table 1.10, columns 3 and 4). From a statistical viewpoint, they further improve the fit of the regression, showing a potential case of omission of relevant variables when they are excluded.

Adoption of auditing schemes (CERTIF, including EMS and ISO schemes, adopted by around 30 per cent of firms) is positively associated with productivity. Significance is shown to be driven by ISO 14000, when we disentangle the total variable using specific dummies. This is somewhat counter-intuitive given that EMS is a stronger strategy in innovation terms. However, this evidence may be consistent with EMS being less frequent than ISO in SME. Also, EMS is strictly aligned to output process–product environmental innovations (Mazzanti and Zoboli 2009), which here show no connections to productivity, confirming our evidence for non-environmental process and product innovations. Although the lack of significance for either type of technological innovation may be due to the short lag between adoption and when we measure productivity performance, our evidence shows stronger impacts for (eco-)organizational innovations. Alternatively, we can assume that organizational innovations tend to shown their influence on labour productivity more quickly than technological innovation effects. ISO 14000 obscures the role of INNO_ORG, although within the general index we find that different elements exert different effects.

For environmental R&D (expenses per employee – Table 1.10, column 5), a factor seldom investigated in empirical studies (and a major result of this analysis) is that its inclusion causes another sharp increase in the fit related to the R^2 and F tests. The R&D variable is highly significant at 5 per cent when corrected for heteroskedasticity, a plausible result given its distribution; it also does not affect the significance of the other covariates. If we consider auditing and R&D jointly, the significance of the latter is not reduced and environmental strategies appear to be key to firm performance, with the already noted relative higher strength of eco-organizational innovations. Thus, both eco-innovation levers (technological and organizational) play a role in determining firm productivity. For standard innovation, we do not find a significant effect for the adoption of eco-innovation (process or product). This may be due to the lag before their adoption exerts any effects, or the need to investigate 'intensity' of innovations (e.g. expenses, radicalness), rather than the adoption.

The relationship between R&D and productivity is a sign that environmental investments do not undermine firm performance even in the short run, and that also their effects are significant. They might be targeted to increasing efficiency at a general level of reasoning, and to generating partially appropriable rents.[11] The significance of the role played by R&D as a technological driver of performance points to the potential of innovation to increase firm performance.

We note that it is eco-R&D expenditures per employee, and not just the presence of R&D activity in the firm (measured and tested by a dummy, not shown and not significant), that is significant. This suggests that it is the intensity rather than the simple adoption of innovation that matters. This may explain why some other output innovation factors that are captured by dummies, ultimately are not significant. Intensity is often difficult to elicit and capture, especially in relation to intangible and non-monetary factors. Further research should focus on the differences between radical and incremental innovation and between intensive and less intensive efforts. Monetizing expenditures may be one way to ground the analysis in more specific and concrete figures.

Finally, the 'policy proxies' we include produce mixed evidence. On the one hand, number of years of waste policies (WASPOL) (with the exception of emissions-related policies) in a given firm seems to reduce performance. This could be because emissions policies are more recent and to control emissions, firms tend first to introduce end-of-pipe technologies rather than fully restructure their processes, an option that in the short run is likely to have a greater negative effect on productivity than the application of end-of-pipe solutions. End-of-pipe strategies are less likely to be feasible for waste, which generally needs a structural approach. It should also be stressed that, while waste policies were implemented in the early and then the late 1990s (packaging waste EU directives), emissions policies are a more recent development, and SME are often not subject to emissions taxes or tradable permit schemes at either EU or Italian level.

We note that the variable capturing current environmental expenses/costs per employee (COSTN) is not significant in any of the regressions. This goes against the mainstream hypothesis that environmental policy costs reduce productivity, as discussed in the literature survey. In combination with other evidence, this would seem to corroborate a sort of Porter reasoning that competitive strategies and investment in innovation assets, including environmental strategies and environmental innovation, if not considered in the very short run, may increase or at least not depress firm performance. We find no evidence of negative effects on performance; however, we do find evidence of non-significant impacts for some of the factors tested, which we assume may be due to the short time lag in our model. Future analyses could test whether the evidence on productivity is sensitive to a longer lag.

Conclusions

We summarize the evidence for the three levels of analysis: drivers of environmental innovations; role of complementarity; and effects of innovation on economic performance.

First, our results provide new evidence on the driving forces of environmental innovation in complex and evolving industrial systems, and open new research directions. The investigation of policy-related effects is a major value added in research at firm and industry levels, with its specific focus on the interplay with endogenous, firm-based strategic decisions related to environmental and other techno-organizational innovation. The role of industrial relations has been overlooked in studies of environmental innovation. Future research should examine the interconnections between environmental and non-environmental innovation, highlighting possible complementarities.

Second, from a policy perspective, our results show that the main driving forces relate to techno-organizational innovation, R&D and the relationships between the firm and its local/regional social partners, such as unions, other firms, universities and research centres. The mere effect of policy impacting on costs is probably not the most effective way of inducing innovation. Policy intervention should focus on and trigger the multiple factors that favour innovation.

Third, in terms of complementarity, although our analysis overall supports the hypothesis of correlation/complementarity between drivers, we can find cases where drivers, such as policy-related costs, R&D, networking and auditing, are not complementary. Complementarity supports innovative dynamics, but cannot be considered the panacea for complex situations with multiple externalities.

The implications of this are that if complementarity holds, this points to the need for policy integration (innovation, certification and strict environmental policy, etc.) within the more circumscribed environmental field or more widely to include environmental aims in other areas. Complementarity generates increasing returns from the implementation of two or more managerial and/or policy efforts. Our evidence generally does not support strong complementarity. Nevertheless, complementarity in a non-strict form is often shown.

Substitutability may signify that policy efforts, or managerial and policy efforts are in conflict and are generating negative spillovers. This highlights the relevance of complementarity assessments at policy and management levels.

Our analysis suggests that in terms of firm management, complementarity among technological and organizational elements increases returns, although this is dependent on the types of environmental innovation and drivers focused on. Thus, complementarity, although important from a policy and managerial perspective, is probably not the solution to the complexity of innovation dynamics.

Finally, we provide unique evidence on SME labour productivity drivers based on our survey and balance sheet data. The drivers we analyse concern techno-organizational and environmental innovations, such as product/process innovations, EMS/ISO, eco-R&D, training, and idiosyncratic local factors including industrial relations and environmental policies, and bring together standard innovation strategies and specific environmental strategies. This produces good statistical performance from our empirical model.

Our results show that the main drivers of standard innovations are training and organizational factors. That control structural factors do not exert effects on performance (apart from a negative effect for chemical firms) is a sign that innovation strategies are more relevant than exogenous factors, such as size and sector of production, at least in the medium term.

The role of technological innovation is shown to be less relevant, signalling possible weakness in terms of innovation intensity and/or longer time span for such techno-innovation in determining real productivity effects. However, this relative ranking confirms the results of previous investigations on the same firms.

Environmental strategies, specifically auditing schemes and especially ISO 14000, and environmental R&D expenditure, appear to improve the fit of the model, and impact positively on productivity. This is a key result of our analysis. Eco-R&D, which is tested here, but is often overlooked (the Community Innovation Survey which will include eco-innovation does not include eco-R&D), encompasses the role of technological innovation. It should be noted also that current expenses associated with environmental costs do not seem to exert a negative effect on performance, as might have been expected. There are some criticalities related to waste policy, but not emissions policy. Comprehensive innovation strategies related to SME seem to impact on firm performance, reducing or even eliminating trade-offs between standard business actions and strategies encompassing environmental objectives. In addition, local idiosyncratic factors, such as participation and the involvement of unions and employees in managerial decisions related to innovation activity, are positively correlated to productivity. Environmental innovation seems to be embedded in the set of innovations firms try to exploit to enhance medium-long run competitiveness, and in the short run, environmental strategies do not show a negative impact.

Future analyses could improve our understanding by focusing on medium-long run lags between the time that an innovation is implemented and when performance is measured. Analyses of complementarities or trade-offs between different eco- and non-environmental innovations, and within categories, would be other rewarding research directions. Finally, the construction of panel datasets derivable from consequential surveys on the same firms would provide value added from a methodological point of view and highlight the role of the dynamics considered here by investigating different lags between innovation and performances.

Notes

1 Works that deal with the impact of regulation on firms' environmental performance, and the effects of environmental innovation on firm performance, including employment, include, among many others, Ziegler and Rennings (2004), Rennings *et al.* (2003, 2004, 2006), Doonan *et al.* (2005), Pfeiffer and Rennings (1999), Cole *et al.* (2005), Konar and Cohen (2001), Ziegler *et al.* (2007), Cainelli *et al.* (2007), and Earnhart and Lizal (2006, 2008). Mazzanti and Zoboli (2008, 2009) provide a more extensive survey.

2 The taxonomy of environmental realms is largely consistent with recent OECD studies (Darnall *et al.* 2005), and seminal conceptual papers (Rennings 2000; Kemp 2000).

3 Specified either in binary forms (specific innovation realms) or as a synthetic index. The dependent variables are denoted: INNO-EM (adoption of process/product environmental innovation related to emissions); INNO-WA (adoption of process/product environmental innovation related to waste); INNO-EN (adoption of process/product environmental innovation related to energy inputs); and INNO-TOT (synthetic index of the adoption of the four environmental innovations).

4 The regressions for materials innovations are not shown since they show a poor fit.

5 The analysis is complex as it involves more than two factors. We refer to Mancinelli and Mazzanti (2009), Milgrom and Roberts (1995) and Mohnen and Roller (2005) for theoretical discussions and applications.

6 These are the possible states of the world, 1 meaning presence and 0 meaning absence of the factor in a defined situation.

7 Non-strict complementarity may be associated with constant returns to scale, while strict complementarity points directly to an effective situation of increasing returns with regard to the analysed factors. This is plausible: complementarity factors may be aimed at maintaining current innovation dynamics and, thus, performance. This, along with increasing returns, should be considered a successful outcome.

8 It also means that, e.g., R&D subsidies and environmental taxes are probably (and plausibly) conflicting factors in this field.

9 Due to the rich data from our three information sources, careful analysis of the correlation matrix shows that high correlation is not a major flaw in this context

10 The indirect role of good industrial relations in enhancing productivity through support for innovation and HPWP in firms, when included in a regression with structural controls as covariates, is not a significant driver of many innovation proxies (e.g.

INNO_ORG, INNOTECH, the proxies for process or product innovation adoption, INNOPROC, RSN, CERTIF).

11 We elicited information on R&D expenditure by asking what share of firm turnover is devoted to environmental R&D, and translating these figures to amount per employee (using official turnover data).

References

Addison, J., Siebert, S., Wagner, J. and Wei, X. (2000) 'Worker participation and firm performance: Evidence from Germany and Britain', *British Journal of Industrial Relations,* 38: 7–48.

Antonioli, D. (2009) 'Industrial relations, techno-organizational innovation and firm economic performance', *Economia Politica*, 26 21–52.

Antonioli, D. and Mazzanti, M. (2009) 'Techno-organisational strategies, environmental innovations and economic performances. Micro-evidence from a SME-based industrial district', *Journal of Innovation Economics*, 2: 145–68.

Antonioli, D., Mazzanti, M. and Pini, P. (in press) 'Productivity, innovation strategies and industrial relations in SME. Empirical evidence for a local manufacturing system in Northern Italy', *International Review of Applied Economics.*

Arora A. (1996), 'Testing for complementarities in reduced-form regressions: A note', *Economic Letters*, 50: 51–5.

Athey, S. and Stern, S. (1998), 'An empirical framework for testing theories about complementarities in organizational design', *NBER Working Paper 6600*, Cambridge, MA: National Bureau of Economic Research.

Becattini, G. (1990) 'The Marshallian industrial district as a socio-economic notion', in F. Cossentino, F. Pyke and W. Sengenberger (eds), *Local Responses to Global Pressures: The Case of Italy and its Districts, Research series no.103*, Geneva: International Labour Office, pp. 37–51.

Biondi, V., Frey M. and Iraldo, F. (2000), 'Environmental management systems and SMEs', *Greener Management International,* 29: 55–69.

Black, S. and Lynch, L. (2001) 'How to compete: The impact of workplace practices and information technology on productivity', *Review of Economics and Statistics,* 83: 434–45.

Booth, A. (1995) *The Economics of the Trade Union,* Cambridge: Cambridge University Press.

Brunnermeier, S. and Cohen, M. (2003) 'The determinants of environmental innovation in US manufacturing industries', *Journal of Environmental Economics and Management*, 45: 278–93.

Brusco, S. (1982) 'The Emilian model: Productive decentralization and social integration', *Cambridge Journal of Economics*, 6: 235–61.

Brusco, S., Cainelli, G., Forni, F., Franchi, M., Malusardi, A. and Rigetti, R. (1996) 'The evolution of industrial districts in Emilia Romagna', in F. Cossentino, F. Pyke, and W. Sengenberger (eds), *Local Responses to Global Pressures: The Case of Italy and its Districts, Research series no.103*, Geneva: International Labour Office, pp. 17–36.

Brynjolfsson, E., Hitt, L. and Yang, S. (2002) 'Intangible assets: Computers and organizational capital', *Brookings Paper on Economic Activity: Macroeconomics*, 1: 137–99.

Cainelli, G., Mazzanti, M. and Zoboli, R. (2007) 'Environmental innovation strategies and firm performances in services', FEEM Working Paper 104, Milan: FEEM.

Caroli, E. and Van Reenen, J. (2001) 'Skill-biased organizational change? Evidence from a panel of British and French establishments', *Quarterly Journal of Economics,* 116: 1449–92.

Cochran, W. (1977) *Sampling Techniques*, New York: John Wiley and Sons.

Cole, M., Elliott, R. and Shimamoto, K. (2005) Industrial characteristics, Environmental regulations and air pollution: an analysis of the UK manufacturing sector, *Journal of Environmental Economics and Management*, 50: 121–43.

Conti, G. (2005) 'Training, productivity and wages in Italy', *Labour Economics*, 12: 557–76.

Coriat, B. and Dosi, G. (2002) 'The nature and accumulation of organizational competences/ capabilities', *Rivista Brasileira de Inovacao,* 1: 275–326.

Darnall, N., Jolley, J. and Ytterhus, B. (2005) 'Does a facility's environmental performance predict its financial performance?' OECD Working Party on National Environmental Policy, Washington DC, OECD, Paris, June, *mimeo*.

Doonan, J., Lanoie, P. and Laplante, B. (2005) 'Determinants of environmental performance in the Canadian pulp and paper industry: An assessment from inside the industry', *Ecological Economics*, 55: 73–84.

Earnhart, D. and Lizal, L. (2006) 'Effects of ownership and financial status on corporate environmental performance', *Journal of Comparative Economics*, 34: 111–29.

Earnhart, D. and Lizal, L. (2008) 'Exploring the links between financial performance and environmental performance in a transition economy', paper presented at the DIME Conference on Eco-innovation, Sustainability and Policy, University of Bordeaux IV, September.

EC (European Commission) (2002) 'New forms of work organization: The obstacles to wider diffusion', Brussels: EC.

Frondel, M., Horbach, J. and Rennings, K. (2004) 'What triggers environmental management and innovation?', *RWI Discussion Papers no.15*, Essen: RWI.

Galia, F. and Legros, D. (2004) 'Complementarities between obstacles to innovation: Evidence from France', *Research Policy*, 33: 1185–99.

Getzner, M. and Ritt, T. (2006) 'Quantitative and qualitative employment impacts of environmental innovations', *International Journal of Environmental Technology and Management,* 4: 375–99.

Granovetter, M. (1985) 'Economic action and social structure: The problem of embeddedness', *American Journal of Sociology*, 91: 481–93.

Griliches, S. (1979), 'Issues in assessing the contribution of R&D to productivity growth', *Bell Journal of Economics*, 10: 92–116.

Gu, W. and Tang, J. (2004) 'Link between innovation and productivity in Canadian manufacturing industries', *Economics of Innovation and New Technology*, 13, 671–86.

Hall, B. and Mairesse, J. (2006) 'Empirical studies of innovation in the knowledge driven economy', *NBER Working Paper 12320*, Cambridge, MA: National Bureau of Economic Research.

Huselid, M. (1995) 'The impact of human resource management practices on turnover, productivity, and corporate financial performance', *Academy of Management Journal*, 38: 635–72.

Ichniowski, C., Shaw, K. and Prennushi, G. (1997) 'The effect of human resource management practices on productivity: A study of steel finishing lines', *American Economic Review*, 87: 291–313.

IPTS (Institute for Prospective Technological Studies) (2004) *Promoting Environmental Technologies: Sectoral Analyses, Barriers and Measures,* Seville: European Commission.

Jaffe, A. and Palmer, K. (1997) 'Environmental regulation and innovation: A panel data study', *Review of Economics and Statistics*, 79: 610–19.

Jaffe, A., Peterson, S.R., Portney, P.R. and Stavins, R. (1995) 'Environmental regulation and the competitiveness of US manufacturing: What does the evidence tell us?', *Journal of Economic Literature,* 33: 32–63.

Kemp, R. (2000) 'Technology and environmental policy – innovation effects of past policies and suggestions for improvement', in OECD (eds) *OECD Proceedings Innovation and the Environment*, Paris: OECD, pp. 35–61.

Khanna, M. and Anton, W. (2002) 'Corporate environmental management: Regulatory and market-based incentives', *Land Economics*, 78: 539–58.

Konar, S. and Cohen, M. (2001) 'Does the market value environmental performance?', *Review of Economics and Statistics*, 83: 281–9.

Kremp, E. and Mairesse, J. (2004) 'Knowledge management, innovation and productivity: A firm level exploration based on French manufacturing CIS3 data', *NBER Working Paper 10237*, Cambridge, MA: National Bureau of Economic Research.

Krozer, Y. and Nentjes, A. (2006) 'An essay on innovations for sustainable development' *Environmental Sciences,* 3: 163–74.

Laursen, K. and Foss, N.J. (2003) 'New human resource management practices, complementarities and the impact of innovation performance', *Cambridge Journal of Economics*, 27: 243–63.

Laursen, K. and Mahnke, V. (2001) 'Knowledge strategies, firm types, and complementarity in human-resource practices', *Journal of Management and Governance*, 5: 1–27.

Leiponen, A. (2005) 'Skills and innovation' *International Journal of Industrial Organization,* 23: 303–23.

Lynch, L. (2007) 'The adoption and diffusion of organizational innovation: Evidence for the US economy', *IZA Discussion Paper 2819*, Bonn: IZA.

Lyon, T. and Maxwell, J. (2008) 'Corporate social responsibility and the environment: A theoretical perspective', *Review of Environmental Economics and Policy Advance Access*, 2: 240–60.

Machin, S. and Wadhwani, S. (1991) 'The effects of unions on investment and innovation: Evidence from Wirs', *The Economic Journal,* 101: 324–30.

Mancinelli, S. and Mazzanti, M. (2009) 'Innovation, networking and complementarity: Evidence on SME performances for a local economic system in North-Eastern Italy, *Annals of Regional Sciences*, 43: 567–97.

Mazzanti, M. and Zoboli, R. (2008) 'Complementarity, firm strategies and environmental innovations', *Environmental Sciences*, 5: 17–40.

Mazzanti, M. and Zoboli, R. (2009) 'Embedding environmental innovation in local production systems: SME strategies, networking and industrial relations', *International Review of Applied Economics*, 23: 169–95.

Mazzanti, M., Pini, P. and Tortia, E. (2006) 'Organisational innovations, industrial relations and firm performance. The Emilia Romagna food sector', *Journal of Socio-Economics,* 35: 123–41.

McKeiver, C. and Gadanne, D. (2005) 'Environmental management systems in small and medium businesses', *International Small Business Journal*, 23: 513–37.

Menezes-Filho, N. and Van Reenen, J. (2003) 'Unions and innovation: A survey of the theory and empirical evidence', in J. Addison and C. Schnabel (eds), *International Handbook of Trade Unions,* Cheltenham: Edward Elgar, pp. 293–334.

Menezes-Filho, N., Ulph, D. and Van Reenen, J. (1998) 'R&D and unionism: Comparative evidence from British companies and establishments' *Industrial and Labour Relations Review*, 52: 45–63.

Metcalf, D. (2003) 'Unions and productivity, financial performance and investment: International evidence', in J. Addison and C. Schnabel (eds), *International Handbook of Trade Unions,* Cheltenham: Edward Elgar, pp. 118–71.

Michie, J. and Sheehan, M. (2005) 'Business strategy, human resources, labour market flexibility and competitive advantage', *International Journal of Human Resource Management*, 16: 445–64.

Milgrom, P. and Roberts, J. (1995) 'Complementarities and fit strategy, structure, and organizational change in manufacturing', *Journal of Accounting Economics*, 19: 179–208.

Mohnen, P. and Roller, L.H. (2005) 'Complementarities in innovation policy', *European Economic Review*, 49: 1431–50.

Montini, A. and Zoboli, R. (2004) 'Environmental impact and innovation in industrial districts', in G. Cainelli and R. Zoboli (eds), *The Evolution of Industrial Districts*, New York: Springer Verlag, pp. 272–315.

OECD (2005) *Oslo Manual. Guidelines for Collecting and Interpreting Innovation Data*, Paris: OECD.

Pencavel, J. (2003) 'The surprising retreat of Union Britain' NBER Working Paper 9564, Cambridge, MA: National Bureau of Economic Research.

Pfeiffer, F. and Rennings, K. (1999) 'Employment impacts of cleaner production', *Discussion paper no.46,* Centre for European Economic Research, Mannheim: ZEW.

Pindyck, R. and Rubinfeld, D.L. (1991) *Econometric Models and Economic Forecasts,* New York: McGraw-Hill.

Porter, M. and Van der Linde, C. (1995) 'Toward a new conception of the environment–competitiveness relationship', *Journal of Economic Perspectives*, 9: 97–118.

Portney, P. (2008) 'The (not so) new corporate social responsibility: An empirical perspective', *Review of Environmental Economics and Policy Advance Access*, 2: 261–75.

Reinhardt, F., Stavins, R. and Vietor, R. (2008) 'Corporate social responsibility through an economic lens', *Review of Environmental Economics and Policy Advance Access*, 2: 219–39.

Rennings, K. (2000) 'Redefining innovation – eco innovation research and the contribution from ecological economics', *Ecological Economics*, 32: 319–32.

Rennings, K., Ziegler, A. and Zwick, T. (2004) 'The effect of environmental innovations on employment changes: an econometric analysis', *Business Strategy and the Environment,* 13: 174–87.

Rennings, K., Ziegler, A., Ankele, K. and Hoffmann, E. (2006) 'The influence of different characteristics of the EU environmental management and auditing scheme on technical environmental innovations and economic performance', *Ecological Economics,* 57: 45–59.

Rennings, K., Ziegler, A., Ankele, K., Hoffmann, E. and Nill J. (2003) 'The influence of the EU environmental management and auditing schemes on environmental innovations and competitiveness in Germany. An analysis on the basis of case studies and a large scale survey', *ZEW Discussion Paper 03-14*, Mannheim: ZEW.

Teece, D. (1996) 'Firm organization, industrial structure and technological innovation', *Journal of Economic Behavior and Organization*, 31: 193–224.

Wooldridge, J. (2002) *Econometrics Analysis of Cross-section and Panel Data*, Cambridge, MA: MIT Press.

Ziegler, A. and Rennings, K. (2004) 'Determinants of environmental innovations in Germany: Do organizational measures matter? A discrete choice analysis at the firm level', *ZEW Discussion Paper 04-30*, Mannheim: ZEW.

Ziegler, A., Schroder, M. and Rennings, K. (2007) 'The effect of environmental and social performance on the stock performance of European Corporations', *Environmental and Resource Economics*, 37: 661–80.

Zwick, T. (2004) 'Employee participation and productivity', *Labour Economics,* 11: 715–40.

2 The dynamic relationship between emissions and manufacturing firms' growth

Giulio Cainelli, Massimiliano Mazzanti and Roberto Zoboli

Introduction

Within the studies on the interrelations between environmental and economic performances of firms, there are currently three research directions of higher added value: (i) the effects of environmental performance and innovations on economic performances, given the relatively wider space the drivers of environmental innovations have occupied (Mazzanti and Zoboli 2008); (ii) an increasing attention to the dynamics of relationships in the short run and medium/long run; (iii) extending at micro and meso levels the analyses from manufacturing to other industries, as recently proposed by Cainelli *et al.* (in press) and Mazzanti and Zoboli (2009a).

This chapter aims at providing a contribution on points (i) and (ii) above. It focuses on manufacturing industry; nevertheless, it uses a very large data set compared with predominant survey-based analyses relying on often small-scale samples. It does not suffer from cross-sectional biases, which often plague *ad hoc* survey data which rarely escape the cross-section trap, if not by repeated surveys over time, since it models a dynamic relationship between *emissions* and firm-level economic growth.

More specifically, this chapter investigates the empirical link between firms' economic growth and emissions, using a very large data set of more than 60,000 Italian manufacturing firms. Economic data refer to the period 2000–2004. In order to circumvent the unavailability of data on environmental performance indicators at this level of microeconomic detail, we use a measure of environmental performance by exploiting sector-based NAMEA data, from which we recover firm-based data. The Italian NAMEA provides detailed data at two-digit level on main emissions, value added and employees over 1990–2003. Taking data for the periods 1990–1999 in order to impose a lag structure to the modelled relationship, we reconstruct emission per employees ratios at firm level by using sector coefficients of emissions per unit of labour, as a pragmatic and only available proxy of environmental performance from official data at firm level.

Related literature

A first main stream of research deals with the drivers of eco-innovation strategies. We here refer for brevity to Chapter 1 of the book.

A second stream of research is focused on environmental innovation and (its) employment effects. The main contributions in this stream include Rennings and Zwick (2001), Rennings *et al.* (2001), Pfeiffer and Rennings (1999). What is relevant to our study is the main hypothesis that increasing environmental performance by environmental innovations strengthens competitiveness and the firm performance, with or without policy stimulus. An ancillary hypothesis is that eco-efficiency investments require higher amounts of labour. The reasoning is that, on the one hand, product innovation spurs employment since it creates new demand, while, on the other hand, process innovations decrease employment since they are usually labour saving. Some employment compensation may occur as a result of indirect price/market driven effects. It should be noted that this is a two-stage process in which first the firm decides whether or not to invest in innovation, and second optimizes the volume of labour following the innovation process.

Rennings and Zwick (2001) is based on a sample of eco-innovative firms for five EU countries, in the manufacturing and service sectors. This is a rather unique study which provides evidence related to manufacturing but also includes some evidence concerning eco-innovations in the service sector. They find that in most firms employment does not change as a consequence of innovation, but this may be due to the limited period covered by the survey. Econometric results show that, apart from some effects registered for product innovations, eco-innovation typologies do not influence the level of employment, though as expected (Caroli and van Reenen 2001), according to their evidence environmentally oriented innovations seem to lead to a skills bias effect. Also, end-of-pipe innovations are related to a higher probability of job losses, while innovations in recycling have a positive effect on employment.

Employment effects may thus be unevenly distributed, with strong negative effects from environmental strategies/policies on low skills intensive industries and potentially positive effects on other industries. It could also be argued that product and process eco-innovation strategies may bring about (potentially negative) net effects on employment, attributable to a destruction of the low skilled labour force (administrative staff) and a creation of high skilled positions (R&D).

Third, there is a literature that has focused on the various relationships between eco-innovation, environmental performance and economic performance.[1] Konar and Cohen (2001) investigated the effect on firm market performance of tangible and intangible assets, including two environmental performance-related elements as explanatory factors. Cohen *et al.* (1997) also analysed the relationship between environmental and financial performances. Overall, these authors found that investing in a green portfolio did not incur a penalty and even produced positive returns.

Less recent works by Gray and Shadbegian (1995) used total factor productivity and growth rates for plants over 1979–1990 as performance indicators to test the impact of environmental regulation and pollution abatement expenditures. They found that $1 more expenditure on abatement is associated with more than $1 worth of productivity losses. They found that, when analysing variation over time or growth rates, the relationship between abatement costs and productivity is not significant. Greenstone (2001) estimates the effects of environmental regulations, using data for 1.75 million observations of plants in the 1967–1987 US censuses of manufacturers. Environmental regulations negatively affect growth in employment, output and capital shipments.

Finally, we would point to recent EU-based studies that focus on the (short-term) effects of environmental strategies on the stock performances of corporations, using standard cross-section/panel approaches (Ziegler *et al.* 2008) and event studies that analyse whether there are exogenous unexpected policy effects on the short-term performance of environmentally minded firms. The latter are criticized for their intrinsic very short-term focus. Although valuable, and based on official data sets, we believe that the value of evidence focusing on stock market performance is limited since the majority of firms, especially in Italy, are medium or small sized, and do not appear in stock-market data. Innovation dynamics are close to productivity trends which, in the end, are the main engines of firm performance.

Our analysis is embedded in the third research directions, focusing on the effects of environmental (emission) intensity on firms' economic performance, here measured by firms' turnover growth. The possible contribution of our work stems from: the focus on a dynamic perspective, the use of real firm performance indicators, the lagged structure of the data set and the very large number of firms. The main 'limit' is the exploitation of sector data instead of micro data for proxing environmental emissions, depending on the general unavailability of real environmental performance measures at the microeconomic level, for a large sample of firms in Italy. This is confirmed by the recent paper by Arimura *et al.* (2008) who study the ISO/EMS (voluntary agreements) effects on environmental performances on the basis of an OECD survey on environmental performance of firms. It is striking that the dependent variable 'environmental performances' is constructed by means of self-reported data using an ordered format (no change, decrease or increase in emissions). Thus, though the first best would to use real environmental performance data at the firm level (challenge for future research), and the exploitation of self-reported survey data is a plausible second best, the (panel) nature of NAMEA data disaggregated at two-digit level offer a robust framework that hybridize firm and sector data to analyse economic–environment relationship.[2]

Conceptual framework, data and methodology

We now shortly specify the main research hypotheses we are testing, that arise from the above discussion. We refer to Mazzanti and Zoboli (2009a,b) and Cainelli *et al.* (2008) for an in-depth discussion of the background conceptual framework.

> **[H.1]** The sign of the dynamic relationship between emissions at time $t-1$ and economic performance specified as firm growth at time t, is investigated by using two different time lags: emissions per unit of labour averaged over 1992–1994 and over 1995–1999.

We expect that the closer the lag is the more probable that environmental performance and economic performances are characterized by a trade-off rather than complementary dynamics. Disentangled dynamics (statistical insignificance) are also a possibility if economic performance is independent of environmental elements.

> **[H.2]** The shape of the relationship is also investigated by specifying quadratic forms for emissions.

Non-linearity is a real-world feature that is analysed in both mainstream and alternative perspectives, but that assumes special relevance in dynamic scenarios. Where the relationship is non-linear, not only the cross-section (sector) heterogeneity and the lag difference between the two 'productivities' are important issues, but also the level of the environmental performances matters. Critical thresholds may exist, and we do expect non-linear dynamics to represent the real-world situation of many environmental–economic co-dynamics. It is nevertheless difficult to assess *ex ante* which shape is the more likely one, given that empirical evidence is still scarce. If the non-linear relationship assumes a U-shape, this would mean that at low emissions the relationship with economic performance is associated to a complementary content, then a trade-off emerges when the environmental performance worsen. Otherwise, in the presence of an inverted U-shape, the trade-off between economic and environmental performances is mitigated at higher environmental intensity. Further increase in environmental intensity hampers growth and leads to a co-dynamics between the two sides of firm objectives. In other words, it may be that only at quite high, or comparatively high, environmental pressures, firms (sectors) endogenously implement (innovative) strategies or are subject to policies, that help achieving a joint dynamic relationship between environmental and economic performances.

Summing up, we carry out various tests on whether environmental performance, deriving from innovation and structural changes leading to emission reduction is, following technology-based and externality-based complementarities relationships (Mazzanti and Zoboli 2008), positively associated with firms' growth. The hypothesis we implicitly assume in the empirical model is that environmental impacts (environmental performance) are also dependent on the core dynamics of innovation, driven by structural factors (e.g. firm size),.policy levers, and idiosyncratic strategic factors (R&D). In other words, we explicitly test whether the two objectives of the firm: environmental and economic performances, are disentangled (no significant correlation), positively related (correlation/complementarity between the two), or negatively correlated (substitution or trade-off framework).

The eventual complementarity may be opposed to the substitution hypothesis which often derives from a usual neoclassic reasoning, which tends to hide the possibility that firms adopt environmental innovation in a non-policy scenario. In fact, if the firm is optimizing resource allocation in production (before environmental regulations), any additional abatement cost or innovation cost deriving from policy enforcement leads, at least in the short run, to an equal reduction in productivity.

The data set used in this analysis was drawn from AIDA: a commercial database collected by Bureau Van Dijck. This large data set of Italian joint stock companies reports balance sheet data such as sales, value added, number of employees, labour cost, technical assets, etc. Using this statistical source, for the period 2000–2004, we built a panel composed of more than 61,000 Italian manufacturing firms. For all these firms we have the following information: (i) the industry in which they operate; (ii) the geographic location; (iii) a size variable measured through the number of employees over the period 2000–2004. In addition, we have, for a sub-sample of these firms, two more pieces of information: i.e., sales and age. This second sample is composed of 36,312 firms. We use these two data sets in relation to the stage of the Heckman procedure we consider below. Specifically, the first larger data set is used to estimate the selection equation where the dependent variable is constituted by a dummy variable that takes the value 1 whether all information on sales and age are available for the period 2000–2004 and the value 0 otherwise. The second smaller data set is used to estimate the growth equations which allow us to investigate the empirical link between environmental and economic performances, respectively emission per employee and turnover. In Tables 2.1 to 2.3 we report the distribution of firms and employees referred to the second data set by geographic areas (Table 2.1), employee classes (Table 2.2) and industry (Table 2.3).

A potential problem with these kinds of samples is that firms are not randomly chosen (Cingano and Schivardi 2004). However, a comparison with the whole population in terms of frequency distribution both by industry and by geographical areas shows that our sample is not far from being representative.

Environmental performances are taken from the NAMEA source. The Italian NAMEA provides detailed data at two-digit level on main emissions, value added and employees over 1990–2007.

Taking data for the period 1990–1999 in order to impose a lag structure to the modelled relationship, and focusing on some primary environmental performance

Table 2.1 Distribution of sample firms by geographic areas

	Firms		*Employees*	
	N	*%*	*N*	*%*
North-West	15,164	41.8	1,182,367	50.3
North-East	11,214	30.9	702,696	29.9
Centre	6,968	19.2	355,228	15.1
South	2,966	8.2	110,993	4.7
Total	36,312	100.0	2,351,284	100.0

Table 2.2 Distribution of sample firms by employee classes

	Firms		*Employees*	
	N	%	*N*	%
0–19	15,607	43.0	169,333	7.2
20–49	12,260	33.8	377,556	16.1
50–249	7,149	19.7	733,364	31.2
>249	1,296	3.6	1,071,032	45.6
Total	36,312	100.0	2,351,284	100.0

Table 2.3 Distribution of sample firms by industry

	Firms		*Employees*	
	N	%	*N*	%
DA	2,584	7.1	158,617	6.7
DB	3,976	10.9	235,098	10.0
DC	1,398	3.8	54,027	2.3
DD	903	2.5	30,123	1.3
DE	2,306	6.4	118,442	5.0
DF	127	0.3	14,069	0.6
DG	1,633	4.5	189,105	8.0
DH	2,003	5.5	118,953	5.1
DI	2,179	6.0	143,185	6.1
DJ	6,895	19.0	365,674	15.6
DK	5,308	14.6	365,321	15.5
DL	3,205	8.8	289,120	12.3
DM	895	2.5	151,985	6.5
DN	2,900	8.0	117,564	5.0
Total	36,312	100.0	2,351,284	100.0

measures of industries (tons of emission for CO_2, SO_x, NO_x per unit of labour, full-time equivalent jobs (FTEJ)), we reconstruct *environmental technical efficiency* ratios at firm level by using sector-level coefficients of emission/FTEJ, i.e. the only available proxy for environmental performance at sector level, and our AIDA data on firm employees. This is the only plausible way to recover firm-level data on environmental performances in order to exploit the very rich information contained respectively in the Italian NAMEA and AIDA.

We set up average emission/employee for the two periods 1992–1994 and 1995–1999 for assessing relationships at different lag distances between environmental performances and firms' growth. Years 1990 and 1991 were in the end discarded, given problems with sector comparability over time within different NAMEA. Average values are taken both for testing the technical efficiency effects using

Table 2.4 Descriptive analysis of variables

Variable and acronym	*Description*	*Mean*	*Std. Dev.*	*Min.*	*Max.*
Y	Turnover growth 2000–2004	0.120	0.590	–6.689	10.137
Turnover 2000	Value of sales in 2000	15.250	1.208	5.860	23.081
Age	Age of the firm (years)	3.089	0.496	1.386	4.955
CO_2	Environmental performance (CO_2 on FTEJ, average value 1995–999)	12.779	1.655	7.826	21.658
NO_x	Environmental performance (NO_x on FTEJ, average value 1995–1999)	6.640	1.489	2.079	15.797
SO_x	Environmental performance (SO_x on FTEJ, average value 1995–1999)	6.184	1.776	0.693	16.781

Note: All variables are to be intended in logs.

two different time periods, and since emission data are characterized by yearly-specific volatility that is mitigated and smoothed by taking averages.

Table 2.4 presents summary estimates for the main variables we use in the econometric analysis.

Empirical analyses

Empirical model and methodological issues

The empirical specification used in this analysis is within the established and well-developed literature based on Gibrat's law on proportionate effects. This hypothesis states that the probability of a given proportionate change in size during a specified period of time is the same for all firms in a given industry, regardless of their size at the beginning of the period (Mansfield 1962). Following Evans (1987a,b), we adopt a growth version of this model, specifying the dependent variable as firm size growth and not firm size at time *t*. The independent variable remains size at time $t - 1$. We test this hypothesis for sales/turnover. Although most studies focus on employment as a proxy for size, there are an (increasing) number of investigations on the literature based on other measures of size and performances, from profitability to asset value.[3]

We opt for the Gibrat model for a variety of reasons. In the absence of a panel of data framework, we deal with a hybrid cross-section environment with a lag structure in the empirical model that circumvents endogeneity. Instead of relying on a simple cross-section specification, we prefer to use such a model given that it has shown good performances in previous studies (Cainelli *et al.* in press) and is related to a well-developed and consolidated literature, mainly established in the evolutionary economics environment. Panel analyses may be scope for future (more valuable) research though we remark that our lagged model is specifically aimed at coping with endogeneity that would be a methodological and conceptual issue in a panel world.

According to this literature, it is also relevant to deal with exit/entry flows over the period. Gibrat's law could also be valid for certain defined sub-samples of firms (young, innovative, etc.). From a methodological point of view, this calls for econometric techniques that tackle sample selection bias.

Finally, some recent papers (Lotti *et al.* 2007) argue that while the law may fail on an *ex ante* basis (that is on the total firms) since small and medium sized firms (SME) grow faster, in an *ex post* equilibrium, after the market has cleaned the industry through competition pressures, this law may hold for the core of survivor firms. Short-run and long-run differences in the validity of Gibrat's law may thus occur, and they can be associated with exit/entry flows and the evolution of industry towards a core set of firms. The period of observation is generally not so long as to detect these differences in the short to long run. In any case our study is not primarily focused on testing Gibrat's law, which is the framework in which we test out hypotheses. Nevertheless, our result should be interpreted as biased towards the short term.

The specification we used to empirically test the relationship between emission intensities and firms size growth is (Evans 1987a,b):

$$\Delta \ln\left(Y_{i,t}\right) = \ln\left(Y_{i,2004}\right) - \ln\left(Y_{i,2000}\right) = \alpha_i + \gamma_i \ln\left(Y_{i,2000}\right) + \delta_i \ln\left(age_i\right) + \mathbf{X}_\mathrm{i}'\beta + v_i \tag{2.1}$$

where $Y_{i,2004}$ and $Y_{i,2000}$ are sales of firm i in 2004 and 2000, age_i denotes firm age, $\mathbf{X}_\mathrm{i}$ is a set of variables, including emissions, and finally v_i is the error term with the usual statistical properties.

It is worth noting that this relationship is also investigated by specifying quadratic terms both for sales and age and for emissions indicators. This is done in order to capture potential non-linear effects among these covariates and the dependent variable.

To overcome potential selection bias, we estimated Equation (2.1) using the Heckman two-step procedure (Cainelli *et al.* in press). In the first step we estimate a selection equation where the dependent variable is constituted by a dummy variable taking value one whether all information on sales and age are available for the period 2000–2004 and the value zero otherwise. As explanatory variables of this selection equation we use four geographic (*North_West*, *North_East*, *Centre* and *South*) and four size (*D19*, *D20_49*, *D50_249* and *D250*) dummies. The residuals of this regression were used to construct a selection bias factor, which is equivalent to the inverse Mill's ratio. This factor accounts for the effects of all unmeasured characteristics which are related to the selection variable. The inverse Mill's ratio is introduced as an extra explanatory variable in the second stage of the Heckman procedure, which consists of estimating the growth equation (2.1) using maximum likelihood estimators and the selection bias control factor as an additional independent variable. In this way, we obtain efficient and consistent estimates of the unknown coefficients of the equations. Finally, since emissions data are calculated at firm level by exploiting emissions/employees ratio from the industry level, we estimate standard errors that are robust to *arbitrary cluster correlation*. This procedure allows us to account for clustering in emission intensity data derived from NAMEA (Wooldridge 2003; 2006).

Empirical results

First, we find expected results that show the relevant influence by size, industry and geographic location.

Secondly, we focus on the core aspects of second stage regressions, where the proper Gibrat model is investigated (see Tables 2.5 to 2.7 for the three different categories of emissions).

As far as the regressions without environmental performances (not shown) indicate significant and negative signs attached to age and turnover: firm growth is more likely to be experienced by small firms, an expected plausible result, and by younger firms, again as expected.

The relationship appears to be nevertheless non-linear: quadratic forms show U-shapes for both factors. This may signify that (very) young and (very) old firms are the ones experiencing higher growth over 2000–2004. The same holds regarding the scale/size of firms: low and high turnovers per employee are associated to higher growths. All in all, non-linearity is empirically relevant. Those outcomes do not change when we include environmental performance indicators (see below).

As a third step, regarding the core analyses of environmental–economic relationships, we first note that linear forms in Tables 2.5 to 2.7 seem to support a positive and significant coefficient for all the three emission categories. Thus, recalling hypothesis [H1], we would be in front of a (usual) trade-off between different kinds of performance: the higher the emissions per employee (over 1995–1999), the higher is firm growth (over 2000–2004). Firm growth thus appears 'not constrained' by bad environmental performances; on the reversal side, more environmentally efficient firms (perhaps belonging to the greener sectors) do not gain the payoffs of 'greener investments'.

Motivations may be many and various. In part this evidence may be linked to weak policy pressures on more polluting firms, mainly when focusing on a global externality like CO_2 which has been regulated from early-2000s on through the IPPC directive and the emission trading (EU ETS) scheme for CO_2. In any case we do not find here evidence of proactive co-dynamics between environmental and economic realms. A positive sign in the estimated coefficient here means

Table 2.5 The impact of environmental performance on firms' growth: estimates (CO_2)

	Heckit[(a)]		*Heckit*[(a)]	
Dep. Var: $\Delta_4 \ln(Y_{i,t})$	*Coefficient*	*t values*	*Coefficient*	*t values*
Selection equation	*[1]*		*[2]*	
North_West	0.663**	8.05	0.067**	8.28
North_East	0.032**	2.38	0.033**	2.42
Centre	Ref.	Ref.	Ref.	Ref.
South	–0.147**	–6.59	–0.150**	–6.84
D19	Ref.	Ref.	Ref.	Ref.
D20_49	0.569**	20.11	0.563**	19.03
D50_249	0.645**	12.23	0.638**	11.96
D250	0.322**	5.05	0.330**	5.21
Industry dummies (13)	Yes	Yes	Yes	Yes
Second stage equation				
$\ln(turnover_{2000})$	–2.559**	–21.32	–2.639**	–24.69
$\ln(age)$	–0.820**	–14.22	–0.852**	–15.43
$\ln(turnover_{2000})^2$	0.075**	18.08	0.078**	22.02
$\ln(turnover_{2000})^2$	0.108**	12.31	0.112**	13.95
$\ln(co2_{1992\_1999})$	0.536**	2.64	0.253**	2.58
$\ln(co2_{1992\_1999})^2$	...	...	–0.007**	–2.37
Mills lambda	–0.340**	–6.93	–0.330**	–6.60
N. Obs.	61,219		61,219	
Censored Obs.	24,907		24,907	
Uncensored Obs.	36,312		36,312	
Clustering	14 industries		14 industries	
Wald $\chi^2(1)$	54.35		50.44	
Prob>χ^2	0.000		0.000	

Notes: (a) Regressions also include a constant term. Standard errors are robust and adjusted for 14 clusters. ** p values < 0.05, * p values < 0.1.

that increasing environmental performance leads to some economic costs in terms of lower growth. Those are the implied cost of eventual improvements in environmental performances, that should be weighted against market (appropriable) and non-market benefits accruing from such improvements. Alternatively, being less environmentally benign relaxes the constraints to growth.

It is worth noting, nevertheless, that the size of the estimated coefficients is not large, besides that for CO_2: trade-offs are statistically significant, but quite negligible in terms of impacts for SO_x and NO_x. Since also the size of coefficients (economic significance) matters as well as its statistical significance (Ziliak and Mcloskey 2004), environmental performance does not seem to be a primary cost and constraint to growth if compared with other factors affecting firm targets.

Table 2.6 The impact of environmental performance on firms' growth: estimates (NO_x)

	Heckit[(a)]		*Heckit*[(a)]	
Dep. Var: $\Delta_4 \ln(Y_{i,t})$	*Coefficient*	*t values*	*Coefficient*	*t values*
Selection equation	*[1]*		*[2]*	
North_West	0.067**	7.92	0.068**	7.62
North_East	0.033**	2.34	0.034**	2.41
Centre	Ref.	Ref.	Ref.	Ref.
South	–0.150**	–6.37	–0.156**	–6.51
D19	Ref.	Ref.	Ref.	Ref.
D20_49	0.564**	18.21	0.554**	17.64
D50_249	0.634**	11.32	0.619**	11.00
D250	0.308**	4.69	0.317**	4.92
Industry dummies	Yes	Yes	Yes	Yes
Second stage equation				
$\ln(turnover_{2000})$	–2.527**	–22.23	–2.635**	–27.01
$\ln(age)$	–0.820**	–13.89	–0.859**	–15.04
$\ln(turnover_{2000})^2$	0.074**	18.45	0.078**	23.28
$\ln(age)^2$	0.107**	11.86	0.113**	13.15
$\ln(nox_{1992\_1999})$	0.070**	2.40	0.234**	4.78
$\ln(nox_{1992\_1999})^2$	…	…	–0.011**	–4.27
Mills lambda	–0.330**	–6.60	–0.311**	–5.55
N. Obs.	61,219		61,219	
Censored Obs.	24,907		24,907	
Uncensored Obs.	36,312		36,312	
Clustering	14 industries		14 industries	
Wald $\chi^2(1)$	48.18		33.68	
Prob>χ^2	0.000		0.000	

Notes: (a) Regressions also include a constant term. Standard errors are robust and adjusted for 14 clusters. ** p values < 0.05, * p values < 0.1.

The result for CO_2 emissions seems to highlight that, by being related to energy consumption and end-of-pipe solutions for CO_2 emissions being limited, to reduce CO_2 emissions may result in a stronger negative effect on firm growth compared with other air pollutants.

Those outcomes also do not mirror the environmental–economic 'efficiency' co-dynamics found by Mazzanti and Zoboli (2009b) who use panel NAMEA data over 1990–2001 for emissions of value added and value added per employee. Nevertheless, the two studies, though conceptually close, are hardly comparable from an empirical point of view: we here exploit a lagged model instead of a panel, with some differences in the considered period, and the analysis was there on all NAMEA sectors, not just manufacturing. Both studies point out the

Table 2.7 The impact of environmental performance on firms' growth: estimates (SO_x)

	Heckit[(a)]		*Heckit*[(a)]	
Dep. Var: $\Delta_4 \ln(Y_{i,t})$	*Coefficient*	*t values*	*Coefficient*	*t values*
Selection equation	*[1]*		*[2]*	
North_West	0.065**	7.79	0.065**	8.01
North_East	0.032**	2.36	0.032**	2.40
Centre	Ref.	Ref.	Ref.	Ref.
South	–0.144**	–6.40	–0.146**	–6.52
D19	Ref.	Ref.	Ref.	Ref.
D20_49	0.574**	21.31	0.567**	20.17
D50_249	0.654**	12.46	0.646**	12.24
D250	0.332**	5.14	0.334**	5.29
Industry dummies (13)	Yes	Yes	Yes	Yes
Second stage equation				
$\ln(turnover_{2000})$	–2.582**	–20.82	–2.648**	–23.99
$\ln(age)$	–0.818**	–13.12	–0.845**	–14.61
$\ln(turnover_{2000})^2$	0.076**	17.79	0.079**	21.19
$\ln(age)^2$	0.108**	12.15	0.112**	13.14
$\ln(sox_{1992\_1999})$	*0.041***	*2.20*	*0.122***	*2.37*
$\ln(sox_{1992\_1999})^2$	...	...	*–0.006***	*–2.13*
Mills lambda	–0.348**	–7.25	–0.340**	–6.66
N. Obs.	61,219		61,219	
Censored Obs.	24,907		24,907	
Uncensored Obs.	36,312		36,312	
Clustering	14 industries		14 industries	
Wald $\chi^2(1)$	60.00		52.34	
Prob>χ^2	0.000		0.000	

Notes: (a) Regressions also include a constant term. Standard errors are robust and adjusted for 14 clusters. ** *p* values < 0.05, * p-values < 0.1.

importance of non-linear paths characterizing the links and evolution of economic and environmental performances.

Finally, the non-linear (quadratic models in right-hand columns of Tables 2.5–2.7) analyses on the income–environment relationship here envisaged gives us additional insights (Hypothesis H2). In fact, regarding CO_2, a robust inverted U, or bell shaped,[5] endogenous dynamics seem to signal the possibility of experiencing a reduction of growth after a threshold is reached. The income–environment relationship is not similar across the range of environmental performances/efficiency. Very inefficient performances in terms of emissions per unit of labour penalise firms even in the core performances. An explanation among others may be that pollution effects are (fully) externalized up to a point, then, when a threshold is surpassed, bad environmental performances present negative effects on firm performances. A linked argument is that of 'diminishing marginal returns' of the 'input' emission, related to energy use and capital intensity. Returns may also become negative as shown here, even from a private perspective, without accounting for the social cost of emission production.

Putting it under a different perspective, a potential co-dynamics between economic and environmental performances emerges when the relationship is non-linear. The negative sign of the quadratic term means that higher emissions per unit of labour hamper growth potential, and vice versa, thus improvements in environmental performances are associated to higher growth for firms. We remark that the non-linear shape is here to be interpreted, though embedded in a dynamic relationship, on the basis of the cross-section dimension. The weight is thus on cross-firm/sector heterogeneity, not temporal heterogeneity that could be captured in panel analysis (Mazzanti *et al.* 2008).

Much of it may depend on the complementarity between private and public elements of emission abatement as a production activity that could be associated with the same technological dynamics. Environmental innovations often give rise to a dual externality, providing the typical R&D spillovers and also reducing environmental externalities (Jaffe *et al.* 2003; Rennings 2000). Therefore, innovation aimed at reducing environmental impact may spur positive innovation spillovers. This element of complementarity could explain why environmental performance may be linked to turnover dynamics.

Another joint motivation is the issues of rent generation and appropriability as well as complementarity in production between environmental and economic (technological) objectives we discussed above. The latter may depend on the features of the environmental goods we deal with (Bruvoll and Medin 2003; Kotchen 2005; Rubbelke 2003; Loschel and Rubbelke 2005).

We may note that for CO_2 this threshold or turning point of the income environment bell-shaped curve is nevertheless outside the range of observed values for turnover growth over 2000–2004.[6] It is then a signal of a potential future reversal of the trade-off into co-dynamics even for mixed pubic goods, like carbon dioxide, whose abatement benefits are not always fully appropriated by firms.

The situation for NO_x is just slightly different: recalling that the coefficient in the linear form is negligible in size, turnover dynamics appear as being associated to a trade-off for a certain part of the relationship. Then a joint dynamics, as before, could characterize the relationship with *win-win* gains. If we estimate the turning point of the bell curve, we note that the NO_x one is lower with respect to CO_2. This is plausible. Though this threshold is still outside the range of observed values, thus predominating the trade-off between economic and environmental performances, the 'fruits' of a joint dynamics seem here closer (in time), given the higher appropriability of rents from innovations improving the NO_x-related efficiency.

Finally, results for SO_x are similar to those for NO_x. All in all, all three main environmental indicators tell a similar story. The short-to-medium term relationship is characterized more by trade-off between environmental performance and firm growth in opposition to co-evolutionary dynamics that nevertheless appear 'at the horizon' through the non-linear relationships robustly emerging from the exploitation of a correction procedure for cluster correlation. For SO_x the threshold turning point shows the closest value to real observed ones. Co-efficiency dynamics are close to being achieved.[7] Innovation and policy levers, though not directly investigated here, probably drive stronger trajectories of co-efficiency for SO_x, which is the emission that has witnessed the strongest decrease in the last 20 years.[8]

Conclusions

We test the extent to which (past) environmental performance, which is driven by structural features and firm strategic actions, including eventual responses to policies, influences firm's growth. The main added value of the *analysis* is the use of real environmental performances data merged with economic performances indicators for a very large sample of firms in a dynamic perspective. The evidence we provide is basically 'policy-free' for carbon dioxide, since Italy did not experience the policies in the 1990s. Though policy valuation will be an interesting point to be added in the future, evidence on how firms behave without regulatory interventions is also of high interest to understand the relationship between economic and environmental performances, as food for thought for management and policy making.

First, a typical trade-off generally appears to emerge for the three environmental emission categories we analyse here, when focusing on a linear specification of the income–environment relationship: less environmental-oriented productions allow higher degrees of freedom and less constraint for growth. We do not find evidence of current pro-active co-dynamics between environmental and economic realms. Increasing environmental performance leads to some economic costs in terms of lower firm growth, at least for our manufacturing firms and in this short/medium run oriented empirical scenario. Alternatively, being less environmentally benign relax the constraints to growth. However, the size of the estimated coefficients is not large besides CO_2: trade-offs are statistically significant but quite negligible in terms of effective average impacts. For example, for SO_x and NO_x growth is reduced by less than 0.10 per cent for a 1 per cent increase of environmental performance; the percentage moves up to half a point of growth for CO_2 abatements. All in all, then, the 'environmental factor' does not seem to be a primary cost and constraint to growth, if compared with other factors affecting firm targets and firm competitiveness.

In addition, and highly important, non-linear analyses of the relationships, nevertheless, tell us to some respect a different story. Results show that the link between emissions and firm growth is in fact robustly non-linear. Inverted U-shapes appear for all three emissions indicators: this may be a signal that both firm strategies and recent policy efforts are affecting the relationship between environmental performance and economic performance, possibly turning it in the near future from a trade-off to joint complementary/co-dynamics. Though our evidence is of cross-sectional nature, capturing sector heterogeneity more than time dynamics, we might affirm that the potential co-dynamics between environmental and economic performances appears close from the evidence, and more likely to be achieved in the future for NO_x and SO_x. This evidence is plausible with the higher appropriability of (higher) environmental performances deriving from innovation actions of manufacturing firms.

The story we discuss here for Italy in the 1990s is a sort of 'policy free' scenario; it is thus consistent that CO_2 appears as the environmental factor less likely to be associated to *win-win* complementary income–environment dynamics in the future. The endogenous evolution of income and environmental factors linked by the web of firms-based and sector-specific innovation contents may not suffice, as highlighted by other studies, for coping with this externality. The relatively less significant evidence for carbon dioxide also calls for policy advice in favour of more stringent policies for emissions that are characterized by a largest part of public good (not appropriable) content.

Further research based on this *analysis* may be provided in the future by using panel data analysis over 2000–2004, exploiting new updated NAMEA data, and estimating the reverse causal effects (economic performance to environmental performances). In the context of this century, some analysis of policy effects may be introduced, among others those deriving from the European emission trading scheme for carbon dioxide operating since 2005.

Another fruitful direction of research, that could help by-passing the absence of firm emission data for large data sets, would be classifying sectors on the basis of their emission intensity (per value added, per employee). The taxonomy may be exploited for testing sector dummies, as it is usual for instance in the typical Pavitt-like taxonomy regarding innovation. Emission intensity taxonomy may be enriched by a classification by energy intensity and/or policy stringency that can lead to a matrix-based structure of such an environmental sector taxonomy.

Notes

1 Foulon *et al.* (2002), Cole *et al.* (2005), Collins and Harris (2005).
2 Earnhart and Lizal (2006, 2007, 2008).
3 Harris and Trainor (2005), Dunne and Hughes (1994), Delmar *et al.* (2003), Audretsch *et al.* (2004), Del Monte and Papagni (2003), Santarelli *et al.* (2006).
4 In our case, the dependent variable in the first stage takes the value one if all information (including sales and age) on firm *i* are available, and zero otherwise.
5 We remark that the estimation of standard errors, that are robust to arbitrary cluster correlation has improved CO_2 results. The bell shape was just weakly significant without such a correction. Even for NO_x, the quadratic specification was completely insignificant.
6 It is then relevant to calculate the turning point to observe whether this is within the observed range of values. Since it is well within the range, it means that besides a share of firms for which the relationship is not constituted by a trade-off, the dynamic is driven by an opposite trajectory wherein environmental productivity and economic productivity diverge.
7 Mazzanti and Zoboli (2009b) note that in the case of air pollutants, provided their emission can improve as a by-product of innovations in energy efficiency and inter-fuel substitution (i.e. 'ancillary benefits' of reducing GHGs), there may

also be *specific* capital stocks capable of reducing some of them, e.g. end-of-pipe technologies reducing SO_x, and the new plant/equipment may be both more capital intensive and less air emissions intensive compared with GHG intensity. This may also take place because of regulation. It should be noted that prior to the Integrated Pollution Prevention Control (IPPC) directive and Europe Union Emissions Trading Scheme (ETS), GHGs were not regulated directly, whereas most air pollutants have been closely regulated since the 1970s in most countries. It is likely that regulation has been the spur for increasing capital–labour ratio to reduce these pollutants.

8 The results do not change if we use average emissions over 1992–1994 instead of 1995–1999.

References

Arimura, T., Hibiki, A. and Katayama, H. (2008) 'Is a voluntary approach an effective environmental policy instrument? A case for environmental management systems', *Journal of Environmental Economics and Management*, 55: 281–95.

Audretsch, D., Klomp, L., Santarelli, E. and Thurik, R. (2004) 'Gibrat's Law: are the services different?', *Review of Industrial Organisation*, 24: 301–24.

Bruvoll, A. and Medin, H. (2003) 'Factors behind the environmental Kuznets curve: a decomposition of the changes in air pollution', *Environmental and Resource Economics*, 24: 27-48.

Cainelli, G., Mazzanti, M. and Zoboli, R. (2008), 'The dynamic relationship between environmental efficiency and manufacturing firm growth', FEEM Working paper 99, Milan: FEEM.

Cainelli, G., Mazzanti, M. and Zoboli, R. (in press) 'Environmentally-oriented innovative strategies and firm performances in services', *International Review of Applied Economics.*

Caroli, E. and Van Reenen, J. (2001) 'Skill-biased organizational change? Evidence from a panel of British and French establishments', *Quarterly Journal of Economics*, 116: 1449-92.

Cingano, F. and Schivardi, F. (2004) 'Identifying the sources of 'ocal productivity growth,' *Journal of the European Economic Association*, 2(4): 720-742.

Cohen, M., Fenn, S. and Konar, S. (1997) 'Environmental and financial performance: are they related?', *mimeo.*

Cole, M.A., Elliott, R. and Shimamoto K. (2005) 'Industrial characteristics, Environmental regulations and air pollution: an analysis of the UK manufacturing sector', *Journal of Environmental Economics and Management*, 50: 121–43.

Collins, A. and Harris, R. (2005) 'The impact of foreign ownership and efficiency on pollution abatement expenditures by chemical plants: some UK evidence', *Scottish Journal of Political Economy*, 52: 747–768.

Del Monte, A. and Papagni, E. (2003) 'R&D and the growth of firms. An empirical analysis of a panel of Italian firms', *Research Policy*, 32: 1003–14.

Delmar, F., Davidsson, P. and Gartner, W.B. (2003) 'Arriving at the high growth firm', *Journal of Business Venturing*, 18: 189–216.

Dunne, T. and Hughes, P. (1994) 'Age, size, growth and survival: UK companies in the 1980s', *Journal of Industrial Economics*, 42: 115–40.

Earnhart, D. and Lizal, L. (2006) 'Effects of ownership and financial status on corporate environmental performance', *Journal of Comparative Economics*, 34: 111–29.

Earnhart, D. and Lizal, L. (2007) 'Does Better Environmental Performance Affect Revenues, Cost, or Both? Evidence From a Transition Economy', William Davidson Institute Working Papers Series 856, William Davidson Institute, Michigan: University of Michigan Stephen M. Ross Business School.

Earnhart, D. and Lizal, L. (2008) 'Exploring the links between financial performance and environmental performance in a transition economy', paper presented at the DIME international conference on Innovation, Sustainability and Policy, University Montesquieu IV, Bordeaux, 11–13 September 2008.

Evans, D.S. (1987a) 'The relationship between firm growth, size and age: estimates for 100 manufacturing industries', *Journal of Industrial Economics*, 35: 567–81.

Evans, D.S. (1987b) 'Tests of alternative theories of firm growth', *Journal of Political Economy*, 95: 657–74.

Foulon, J., Lanoie, P. and Laplante, B. (2002) 'Incentives for pollution control: Regulation or information?', *Journal of Environmental Economics and Management,* 44: 169–87.

Gray, W. and Shadbegian, R. (1995) 'Pollution, Abatement Costs, Regulation, and Plant-level Productivity', NBER Working Paper 4994, Cambridge, MA: NBER.

Greenstone, M. (2001) 'The Impacts of Environmental Regulations on Industrial Activity: Evidence from the 1970 and 1977 Clean Air Act Amendments and the Census of Manufacturers', NBER Working Paper 8484, Cambridge, MA: NBER.

Harris, R. and Trainor, M. (2005) 'Plant level analyses using the ARD: another look at the Gibrat's Law', *Scottish Journal of Political Economy*, 52: 492–518.

Jaffe, A., Newell, R. and Stavins, R. (2003) 'Technological change and the environment', in Mäler, K.G. and Vincent, J. (eds.), *Handbook of Environmental Economics*, Vol.1, Amsterdam: Elsevier. .

Konar, S. and Cohen, M. (2001) 'Does the market value environmental performance?', *The Review of Economics and Statistics*, 83: 281–9.

Kotchen, M. (2005) 'Impure public goods and the comparative statics of environmentally friendly consumption', *Journal of Environmental Economics and Management,* 49: 281–300.

Loschel, A. and Rubbelke, T.G. (2005) 'Impure public goods and technological Interdependencies', FEEM Working paper 60, Milan: FEEM.

Lotti, F., Santarelli, E. and Vivarelli, M. (2007) 'Defending Gibrat's laws as a long run regularity', IZA Discussion Paper 2744, Bonn: IZA.

Mansfield, K. (1962) 'Entry, Gibrat's law, innovation and the growth of firms', *American Economic Review*, 52: 1023–51.

Mazzanti, M. and Zoboli, R. (2008) 'Complementarity, firm strategies and environmental innovations', *Environmental Sciences*, 5: 17–40.

Mazzanti, M. and Zoboli, R. (2009a) 'Embedding environmental innovation in local production systems: SME strategies, networking and industrial relations', *International Review of Applied Economics*, 23: 169–95.Mazzanti, M. and Zoboli, R. (2009b) 'Environmental efficiency and labour productivity: trade-off or joint dynamics?', *Ecological Economics*, 68: 1182–94.

Mazzanti, M., Montini, A. and Zoboli, R. (2008) 'Environmental Kuznets curve for air pollutant emissions in Italy: evidence from environmental accounts NAMEA panel data', *Economic Systems Research*, 20: 279–305.

Pfeiffer, F. and Rennings, K. (1999) 'Employment impacts of cleaner production', Discussion paper 46, Centre for European Economic Research (ZEW), Mannheim: ZEW.

Rennings K. (2000), Redefining innovation – eco innovation research and the contribution from ecological economics, *Ecological Economics*, 32 (2): 5–17.

Rennings, K. and Zwick, T. (2001) 'The Employment Impact of Cleaner Production on the Firm Level. Empirical Evidence from a Survey in Five European Countries', Discussion paper 8, Centre for European Economic Research (ZEW), Mannheim: ZEW.

Rennings, K., Ziegler, A. and Zwick, T. (2001) 'Employment changes in environmentally innovative firms', Discussion paper 46, Centre for European Economic Research (ZEW), Mannheim: ZEW.

Rubbelke, D. (2003) 'An analysis of differing abatement incentives', *Resource and Energy Economics,* 25: 269–94.

Santarelli, E., Klomp, L. and Thurik, R. (2006) 'Gibrat's Law: an overview of the empirical literature', in Santarelli, E. (ed.), *Entrepreneurship, growth and innovation: the dynamics of firms and industries*, New York: Springer.

Wooldridge, J.M. (2003) 'Cluster-sample methods in applied econometrics', *American Economic Review*, 93: 133–8.

Wooldridge, J.M. (2006) 'Cluster-Sample Methods in Applied Econometrics: An Extended Analysis', *mimeo,* Department of Economics, Michigan: Michigan State University.

Ziegler, A., Schroder, M. and Rennings, K. (2008) 'The effect of environmental and social performance on the stock performance of European corporations', *Environmental & Resource Economics*, 37: 661–80.

Ziliak, S. and Mcloskey, D. (2004) 'Size matters: the standard error of regressions in the American economic review', *Journal of Socio Economics*, 33: 527–46.

3 Environmentally-orientated strategies and firm performance in services

Giulio Cainelli, Massimiliano Mazzanti and Roberto Zoboli

Introduction

Based on its relatively high environmental impact, and high innovation potential, the manufacturing sector is a focus of attention, while the service sector, although representing some 60–70 per cent of GDP in most OECD countries, seems to be of less interest. If it is true that services are relatively more environmentally benign, it is also generally accepted that they are affected to different degrees by Baumol's disease, which reduces their innovation potential, and along a dynamic path potentially undermines increases in environmental and economic efficiency. For example, using NAMEA panel data, Femia and Panfili (2005) show that, from an environmental point of view, services are more efficient than industry, although not by as much as might be expected and with highly branch-specific evidence, if we apply a dynamic lens. The reason for this is perhaps that the service sector induces transformation of matter, even though their 'product' is not directly material. Support for this view emerges from product life cycle and material flow analyses based on input–output frameworks. From a different perspective, Kander (2005) observes that there is good reason to be sceptical about the idea that the transition to a service economy will bring about dematerialization of production and consequent environmental improvements. This is because a shift to a service economy may be illusory in terms of real production, if it is generated by a fall in the price of manufactured goods relative to services, which in turn is caused by more rapid productivity growth in manufacturing than in services.

This chapter makes four contributions to the empirical literature.

First, we provide new (perhaps the first) evidence on how environmental innovation strategies impact on firm economic performance indicators such as employment, turnover, and labour productivity growth in the services sector.

Second, in contrast to other (intrinsically) survey based analyses, we exploit real performance indicators rather than elicited subjective variables.

Third, we use both employment and turnover in order to verify possibly different links to innovation as a driving force: labour productivity, the core indicator for firm economic performance is studied as a ratio of turnover and employment.

Fourth, again unlike most firm-based studies using surveys, we exploit data derived from merging Community Innovation Survey (CIS) data and balance sheet data. We refer to Chapters 1 and 2 of this book for a survey of the related literature.

The empirical and conceptual framework

Dataset

The investigation is based on an original longitudinal dataset built by matching data from two different statistical sources: (i) the second wave of the Italian Community Innovation Survey (CIS II) and (ii) the System of the Enterprise Accounts (SEA). The resulting sample is composed of 773 service firms with 20 or more employees for which a wide set of innovative data for the period 1993–1995, and a selected number of economic indicators such as employment and turnover for the period 1995–1998 are available. We stress that CIS III and IV, even if available, would not be useful for the aim of our study. CIS II is the only one containing information on 'environmental innovation' motivations. Future CIS will start eliciting some questions on adoptions of environmental technologies (product/process), allowing in a first stage only the analysis of drivers of eco-innovation. Then the effects of such innovations on economic performances will be possibly analysed when data are available. This is food for future research using CIS surveys.

Table 3.1 shows that our sample closely mirrors the full CIS II population in terms of percentage of innovative firms in total firms and overall structure. Table 3.2 shows the distribution of sample firms by service sectors and size.

We measure service firms' economic performance using three indicators: (i) rate of growth of employment over the period 1995–1998; (ii) rate of growth of turnover at current prices over the period 1995–1998; and, finally, (iii) rate of growth of labour productivity, measured as the ratio between turnover at current prices and number of employees. As already mentioned these performance indicators are expressed in terms of current prices; thus, they may subject to price change effects.

Great care was taken in the empirical identification of the sector and size dummies which was based on earlier work on these issues (Cainelli *et al.* 2006). Sector dummies were selected to capture sector-specific technological regimes as well as structural differences between sectors in terms of funding and conducting innovation activities. See the Appendix for a more detailed description.

For the reasons behind innovative strategies such as cost reduction, extending the array of services/products supplied, increasing or penetrating in new markets, developing services with lower environmental impact/output and so on, we constructed for each of these ten innovative strategies considered in the CIS II questionnaire a dummy variable, assigning the value one to the responses (d) averagely relevant, (e) very relevant and (f) crucial, and the value zero to (a) not relevant, (b) low relevance and (c) moderately relevant.[1]

Table 3.3 shows the relative shares of strategic motivations for innovation.

Table 3.1 A comparison between CIS II population and the sample

	CIS II population			*Sample*		
Service sectors	*Total firms*	*%*	*% innovating firms*	*Total firms*	*%*	*% innovating firms*
Trade	8,310	43.7	29.3	227	29.4	48.0
Hotels & restaurants	2,186	11.5	19.6	45	5.8	40.0
Transport	2,828	14.9	29.6	230	29.8	47.8
Waste disposal	255	1.3	27.8	19	2.5	31.6
Software & related	972	5.1	54.3	55	7.1	89.1
R&D, engineering, technical consultancy	435	2.3	55.4	37	4.8	75.7
Legal and marketing	677	3.6	34.9	24	3.1	62.5
Security, cleaning, other business services	2,069	10.9	19.3	132	17.1	28.0
Post and telecommunications	55	0.3	10.9	3	0.4	100.0
Financial services	1,237	6.5	61.9	1	0.1	100.0
Total	19,024	100.0	31.3	773	100.0	48.6

Table 3.2 The structure of the sample: service firms by sector and size

	20–99		*100–249*		*250 and more*	
Service sectors	*N*	*%*	*N*	*%*	*N*	*%*
Trade	88	50.0	64	27.2	75	20.7
Hotels and restaurants	6	3.4	14	6.0	25	6.9
Transport	29	16.5	70	29.8	131	36.2
Waste disposal	5	2.8	4	1.7	10	2.8
Software and related	12	6.8	16	6.8	27	7.5
R&D, engineering, technical consultancy	9	5.1	12	5.1	16	4.4
Legal and marketing	6	3.4	7	3.0	11	3.0
Security, cleaning, other business services	21	11.9	47	20.0	64	17.7
Post and telecommunications	0	0.0	0	0.0	3	0.8
Financial services	0	0.0	1	0.4	0	0.0
Total	176	100.0	235	100.0	362	100.0

Table 3.3 Distribution of answers about innovative strategy by aims (%)

Aims	*Not relevant*	*Low relevance*	*Moderately relevant*	*Averagely relevant*	*Very relevant*	*Crucial*	*No answer*	*Total*
[1] Modify the array of services/products supplied	25.0	7.4	11.7	18.4	14.6	11.2	11.7	100.0
[2] Extend the array of services/products supplied	18.1	2.9	7.4	13.6	25.3	25.0	7.7	100.0
[3] Develop services with lower environmental impact/output	*45.7*	*9.0*	*6.9*	*10.9*	*7.2*	*7.7*	*12.5*	*100.0*
[4] Maintain current market shares	25.5	5.1	6.1	14.4	18.9	18.4	11.7	100.0
[5] Increase market shares	15.7	2.4	3.2	13.0	25.0	34.0	6.6	100.0
[6] Penetrate new markets	27.4	6.9	8.5	12.5	14.4	19.1	11.2	100.0
[7] Improve production/plant flexibility	17.6	3.2	7.4	18.1	23.1	22.1	8.5	100.0
[8] Reduce production costs	13.3	2.7	6.1	17.3	22.6	30.9	7.2	100.0
[9] Increase the quality of services/products	2.9	1.3	1.9	11.4	25.0	55.1	2.4	100.0
[10] Improve the workforce job related welfare	8.8	5.3	12.2	25.5	21.8	19.4	6.9	100.0
[11] Adapt technologies to currently prevailing ones	13.0	6.4	9.3	22.6	22.1	18.6	8.0	100.0

Empirical model and methodological issues

The empirical specification in this chapter is within the established and well-developed literature based on Gibrat's law on proportionate effects that was already presented in Chapter 2.

This hypothesis states that the probability of a given proportionate change in size during a specified period of time is the same for all firms in a given industry, regardless of their size at the beginning of the period (Mansfield 1962).

Following Evans (1987a,b), we adopt a growth version of this model, specifying the dependent variable as firm size growth and not firm size at time t. The independent variable remains size at time $t - 1$. We test this hypothesis for employment, turnover and labour productivity. Although most studies focus on employment as a proxy for size, there are an (increasing) number of investigations on the literature based on other measures of size and performances, from profitability to asset value. We refer to Chapters 1 and 2 for a longer discussion. Our study in any case is not primarily focused on testing Gibrat's law. Nevertheless, any result should be interpreted as biased towards the short to medium term.[2]

The specification we use to empirically test the effects of environmental innovation strategies on firm growth controlling for other firm characteristics and innovative strategies is represented by equation:

$$\Delta_3 \ln\left(Y_{i,t}\right) = \ln\left(Y_{i,1998}\right) - \ln\left(Y_{i,1995}\right) = \alpha_i + \ln\left(Y_{i,1995}\right) + \sum_{j=1}^{10} D\_strategy_{i,j,1993-1995} + \mathbf{X}_i'\beta + v_i \qquad (3.1)$$

where $Y_{i,1998}$ and $Y_{i,1995}$ are performance indicators for firm i in 1998 and 1995, measured either as employment, turnover or labour productivity (measured as the ratio between turnover and employees), $D\_strategy_{i,j,1993-1995}$ is a set of dummy variables, capturing the intensity of each innovative strategy, $\mathbf{X}_i$ is the vector of controls, and ν_i is the error term with the usual statistical properties.

In order to overcome potential selection bias, we estimated Equation (3.1) using the Heckman two-step procedure (Cainelli *et al.* 2006). The first step consists of estimating a probit model of a dummy variable. In our case, the latter takes the value one if the service firm has introduced a technological innovation during the period 1993–1995 and zero otherwise, and is explained by a set of variables available for all firms in the sample (innovative and non-innovative). The covariates used in the first stage are the following: a constant term, three geographic dummies (*North-West*; *North-East* and *Centre*), two size dummies (*D100_249* and *D250*), nine sector dummies and a group dummy (*DGroup*). The residuals of this regression were used to construct a selection bias factor, which is equivalent to the inverse Mill's ratio. This factor accounts for the effects of all unmeasured characteristics which are related to the selection variable. To sum up, the empirical model of reference for the analysis is a model inspired and embedded in the Gibrat's law empirical literature with the emphasis on innovation-like covariates, which tackles the sample bias regarding innovation by setting a two-stage Heckman model, which is usual in this literature (Calvo 2006; Lotti *et al.* 2001; Evans 1987a, 1987b).

Main conceptual links investigated

We here briefly make more explicit and discuss the main conceptual links between innovation and economic performances we test.

The first key issue (Hypothesis 1, H1) is related to the effect of (eco-) innovation strategies on employment growth. Assigning a specific hypothesis to this link is difficult as it is the net effect on employment levels and growth depending on the 'sum' of different positive and negative effects stemming from innovation changes and innovation adoptions. Generally speaking, in the absence of detailed data on the skill content of the workforce, we can expect that value added oriented strategies to impact positively and cost reduction strategies to impact negatively. In terms of the effects of environmentally oriented strategies, the theoretical and empirical literature has highlighted that these will depend both on the kind of innovation adopted (mainly product versus process) and on the type of workforce involved. It is the matching between these two elements that is important for defining eventual net effects. We can say that negative effects on the low-skilled workforce and on levels and growth rates, relatively speaking, are more likely when process innovations are the key strategy of firms in relation to the environment.

Pfeiffer and Rennings (1999) provide a taxonomy of possible effects of eco-innovations, embedded in the wider realms of innovation and employment dynamic relationships specifying hypotheses in relation to product, process, recycling and organizational innovations. Generally speaking, they stress that both positive and negative effects are possible. On the one hand there could be job losses deriving from increases in work productivity linked to the adoption of new technologies. On the other hand, new jobs stemming from increased competitiveness (brought about either by higher cost efficiency and/or by higher market value added potential) could arise. Thus, the type and content of eco-innovation matter: process innovations are likely to cause direct negative effects and product innovations direct positive effects. Nevertheless, compensatory or reinforcing indirect effects, probably occurring in a medium long run scenario, are also likely to exist: process innovations could impact positively on final performance through higher productive efficiencies. Then, higher market shares could impact on employment. Product innovations tend to impact positively even in secondary stages, though the degree of complementarity between old and new products is not irrelevant in assessing employment impacts in equilibrium. According to Pfeiffer and Rennings, and we largely support this view, recycling measures and organizational innovations (EMS) should impact positively, given their labour-intensive content and, in the latter case, value adding processes.

In the second hypothesis, (Hypothesis 2, H2: from eco-innovation strategy to firm economic performance) we can see from the above discussion that the substitution hypothesis which often derives from normal neoclassic reasoning, tends to hide the possibility that firms adopt environmental innovation in a non-policy or BAU scenario.

Nevertheless, even heterodox-minded authors (Rennings *et al.* 2001: 4) state that: ‘environmentally friendly innovation does not necessarily increase the productivity of a firm, however. They may even reduce productivity and require increasing labour inputs per unit because their are often not motivated by cost reduction or increasing sales (both potential positive drivers of productivity), but by compliance with environmental regulations (Cleff and Rennings 1999) and therefore the net effect is unclear’.

Thus, we can conclude that significant (negative, positive) or insignificant signs may be expected. Positive signs should prevail in a long-run scenario, and negative in a short-run effect. As our causal structure is short to medium run, we argue that the results are quite open. In addition, productivity effects stem from the composition of employment and value added (turnover effects). As for profitability effects which depend on the sum of cost and turnover impacts, evaluation of the results should take account of what occurs at the level of both employment and turnover. This is possible given the information we possess.

In both cases, then, the relationships between eco-innovation strategies and firm performances are highly sector specific. We are nevertheless prevented from carrying out analyses on specific sub-sectors, given the insufficient amount of data for carrying out the analysis if we disaggregate them.

Finally, we can see that the early 1990s, and this emerges from the shares of eco-firms, were characterized by low commitment to improving the environment. A quarter of firms, probably leaders and first movers, may have anticipated the market. If short-run effects prevail, and we assume that the market was at that time not yet sufficiently mature to absorb environmental innovation, in terms, say, of higher demand from final consumers or other firms (suppliers, sub-contractors, clients), along vertical and horizontal market relationships, there will be a higher probability of negative effects of eco-strategy on productivity. The current picture may be different, there is no a stricter set of policy levers in place, even in services, and a more developed markets for eco-products at all stages, with a larger share of firms reinforcing the market and the competition in the market in these areas with higher firm awareness and greater firm commitment. The key question revolves around the possibility that firms may adopt some environmental strategies even on an endogenous market-based path.

Empirical evidence

Employment growth effects

Our main finding is of a negative relationship between eco-innovation strategies and firm growth (Table 3.4). In addition we find that larger firms grow less. This latter result is consistent with findings such as Evans (1987a,b), who studied manufacturing industries in a Gibrat’s law framework, taking growth rates as the

Table 3.4 The impact of innovation strategies on firms' employment growth: estimates

	Heckit[(a)]		*Heckit*[(a)]		*Heckit*[(a)]	
Estimation method	*Coefficient*	*t values*	*Coefficient*	*t values*	*Coefficient*	*t values*
Selection equation	*[1]*		*[2]*		*[3]*	
North-West	0.436**	2.62	0.483**	3.13	0.478**	3.05
North-East	0.263	1.52	0.324**	2.04	0.324**	2.02
Centre	0.303*	1.67	0.282*	1.68	0.281*	1.68
South	Ref.	Ref.	Ref.	Ref.	Ref.	Ref.
D20_99	Ref.	Ref.	Ref.	Ref.	Ref.	Ref.
D100_249	0.384**	2.66	0.404**	2.71	0.394**	3.35
D250	0.548	4.06	0.634**	5.11	0.630**	5.59
DTrade	0.585*	1.92	0.651**	2.27	0.655**	2.30
DHotel	0.272	0.76	0.403	1.22	0.419	1.30
DTrasp	0.372	1.24	0.443	1.52	0.420	1.47
DPost	6.860**	12.92	7.362**	18.46	7.299**	8.38
DFinan	6.832**	18.68	7.368**	14.81	7.209**	18.32
DComp	1.851**	4.96	1.952**	6.46	1.966**	6.77
DRDcon	1.254**	3.44	1.449**	4.60	1.421**	4.63
DLegmkt	0.815**	1.98	0.966**	2.82	0.947**	2.77
DOthbus	–0.021	–0.07	0.108	0.36	0.122	0.41
DWaste	Ref.	Ref.	Ref.	Ref.	Ref.	Ref.
DGroup	0.442**	2.47	0.390**	2.86	0.388**	2.87
Second stage equation						
Ln(employees$_{95}$)	–0.068**	–3.20	–0.059**	–3.02	–0.026	–1.57
D_strategy_1	…	…	–0.010	–0.19	–0.002	–0.06

D_strategy_2	…	…	0.026	0.45	0.022	0.39
D_strategy_3	…	…	–0.140**	–3.40	–0.131**	–3.37
D_strategy_4	…	…	–0.117**	–2.39	–0.116	–2.38
D_strategy_5	…	…	0.108*	1.86	0.097*	1.74
D_strategy_6	…	…	0.016	0.40	0.002	0.07
D_strategy_7	…	…	0.124**	2.32	0.125**	2.44
D_strategy_8	…	…	–0.034	–0.57	–0.037	–0.66
D_strategy_9	…	…	0.013	0.14	0.023	0.24
D_strategy_10	…	…	–0.019	–0.34	–0.003	–0.07
D_strategy_11	…	…	0.052	1.23	0.057	1.44
Sector dummies	No		No		Yes	
Size dummies	No		No		Yes	
Mills lambda	–0.025	–0.27	0.21	1.10	–0.033	–0.34
Censored obs.	397		397		397	
Uncensored obs.	376		304		304	
Obs.	773		701		701	
Wald $\chi^2(1)$	10.24		31.08		48.16	
Prob>χ^2	0.0014		0.0019		0.0001	

Notes: (a) The regressions also include a constant term. ** p values < 0.05, * p values < 0.1.

dependent variable, and Nelson and Winter (1982) who speculate on the empirical results, in terms of whether it is plausible that firm growth initially increases with size but then decreases. It is also consistent with new evidence which, on average, tends to reject the law more often compared with earlier studies that tend to confirm it (Lotti *et al.* 2007).

Taken jointly, these outcomes tell us that firm employment growth over 1998–1995 is lower in larger firms and in firms (25 per cent of the total) adopting environmentally oriented innovation strategies. It should be noted that, though ancillary in our analysis, the innovation strategy aimed at preserving market quotas is associated with a negative coefficient on growth, while firms that try to extend market shares through innovations activities appear to grow more in employment terms. A value added oriented strategy, therefore, is more rewarding in terms of employment performance. We note that here and at other levels of analysis that the inclusion in the second stage of size and sector controls does not affect the results.

One plausible interpretation of the significant negative impact of eco-innovation strategies on employment growth is the following. First, it may derive from efficiency improvements (dematerialization processes) that also impact on firm efficiency indicator through reductions in the workforce. Most environmental-intensive processes could be hypothesized to be also labour intensive. A reduction in material and emission flows at the organizational level in the various steps of production and distribution of goods and services could easily be associated with cuts in labour inputs due to technology or services substituted by more efficient structures, processes or completely dematerialized. Such effects could be highly heterogeneous across sectors. Unfortunately, the estimates we carried out on major sub-sectors were not statistically meaningful, probably due to the reduced number of firms per sector. This could be the subject of further research: it may seem trivial but employment effects driven by technological changes may differ from trade to finance to R&D companies.

Nevertheless, and in addition, it would be helpful for future research to disentangle the effects on low- and high-skilled workers. Although it is evident that the net negative effect is an empirical possibility among others, quite contingent on the period and firms observed, it is also likely that it could derive from the destruction of low-skilled employment and the creation of high-skilled jobs, as a consequence of increased environmental awareness and strategy. Most environmental strategies (a counter-example might be the waste management) are accompanied by a reduction in the low-skilled workforce and an increase in highly qualified staff. In terms of numbers, the latter may be impacting less on the net figures, leading to negative impacts.

This is not inconsistent with the evidence from other studies. For example, Pfeiffer and Rennings (2001) assess the net employment effects of technical progress that can be expected from the ongoing transition from end-of-pipe technologies to cleaner production. The main result is that cleaner production leads to the net creation of jobs in more firms than do end-of-pipe technologies. However, eco-innovations like other innovations, tend to require higher qualifications. Thus, the demand for skilled and high-skilled labour rises while the demand for unskilled labour decreases.

Another interpretation of such evidence is possible focusing on the concept of eco-innovation. We can expect that product innovations to be more benign regarding employment effects, since they are linked to the creation of value added, while process integrated innovations (and to a lesser extent end-of-pipe technologies) may destroy the workforce, by substituting technological inputs or reducing the layers/steps in the production activity. Rennings *et al.* (2003) explore the determinants of employment changes due to environmental innovations and show that if the most important environmental innovation is a product or service innovation it has a significantly positive effect on the probability of an increase in employment compared with the probability of no noticeable change. In contrast, if the most important environmental innovation is an end-of-pipe innovation it has a significantly positive influence on employment decrease. Even on a general basis, negative effects (on the low-skilled workforce), relatively speaking, are more likely when process innovations are the key strategy of firms (in relation to the environment). This is compatible with the picture provided by Crespi and Pianta (2008) and Evangelista and Savona (2002, 2003): Italy is a country 'biased' towards process innovation (vs product innovation), one of the factors behind the 'low value added' dynamics of Italian firms. Process innovations and the aforementioned recent dynamics in services could even explain the negative impact on employment growth of eco-innovations.

Turnover growth effects

We note that environmental strategies are associated with a reduced growth in terms of turnover (Table 3.5). This implies that short-run effects underlined by the mainstream literature could be in action. Leaders might find it difficult to reap returns from such strategies for the reasons outlined above, or because the intensity (which we do not observe) is not sufficiently high to cause a change in production efficiency and demand through environmental innovation dynamics. We tend to exclude the possibility of real negative effects that may persist in the long run or in the evolution of markets, since such strategies do not arise from exogenous policy impacts, which are more likely to cause persistent negative effects. We opt for the idea that such negative effects are due to first movers acting in still not mature 'market' and institutional environments, and to short to medium run dynamics, not in equilibrium.

There is also confirmation of a negative and highly significant link between size and size growth, and a positive effect of strategy aimed at improving the labour conditions of workers (improving job-related welfare for the workforce). This at first sight is an unexpected result. We can interpret it from the perspective and the empirical results of recent studies dealing with techno-organizational innovation (such as HRM) and firm performance. There is mounting evidence that firm performance is driven by HRM and innovation factors. Works that have analysed the labour condition effects of such innovations provide ambiguous evidence,

Table 3.5 The impact of innovation strategies on firms' sale growth: estimates

	Heckit[(a)]		*Heckit*[(a)]		*Heckit*[(a)]	
Estimation method	*Coefficient*	*t values*	*Coefficient*	*t values*	*Coefficient*	*t values*
Selection equation	*[1]*		*[2]*		*[3]*	
North-West	0.234*	1.89	0.344**	2.66	0.367**	2.47
North-East	0.189	1.53	0.255**	1.97	0.269*	1.75
Centre	0.154	1.22	0.196	1.48	0.223	1.42
South	Ref.	Ref.	Ref.	Ref.	Ref.	Ref.
D20_99	Ref.	Ref.	Ref.	Ref.	Ref.	Ref.
D100_249	0.347**	3.22	0.320**	3.04	0.465**	4.12
D250	0.538**	4.82	0.551**	5.21	0.699**	6.43
DTrade	0.406**	2.05	0.464**	2.16	0.418*	1.90
DHotel	0.044	0.21	0.190	0.82	0.163	0.70
DTrasp	0.069	0.37	0.153	0.73	0.125	0.59
DPost	7.364**	8.54	8.979**	7.10	9.286**	6.94
DFinan	3.571**	4.42	3.849**	3.82	3.533**	3.81
DComp	1.580**	3.39	1.711**	4.42	1.712**	4.18
DRDcon	0.830**	2.85	1.056	3.74	1.020**	3.69
DLegmkt	0.744**	2.20	0.853**	2.78	0.819**	2.61
DOthbus	–0.208	–1.04	–0.122	–0.55	–0.123	–0.56
DWaste	Ref.	Ref.	Ref.	Ref.	Ref.	Ref.
DGroup	0.091	0.77	0.028	0.25	0.036	0.32
Second stage equation						
Ln($sales_{95}$)	–0.069**	–4.20	–0.080**	–4.37	–0.062**	–3.25
D_strategy_1	…	…	–0.072	–0.76	–0.059	–0.62

D_strategy_2	…	…	0.122	1.15	0.112	1.15
D_strategy_3	…	…	–0.144**	–2.12	–0.133**	–1.96
D_strategy_4	…	…	–0.058	–0.77	–0.060	–0.79
D_strategy_5	…	…	0.083	0.86	0.066	0.68
D_strategy_6	…	…	–0.002	–0.05	–0.015	–0.26
D_strategy_7	…	…	0.053	0.56	0.061	0.63
D_strategy_8	…	…	–0.114	–1.05	–0.111	–1.00
D_strategy_9	…	…	0.085	0.74	0.109	0.96
D_strategy_10	…	…	0.188**	2.64	0.190**	2.69
D_strategy_11	…	…	0.080	1.26	0.088	1.42
Sector dummies	No		No		Yes	
Size dummies	No		No		Yes	
Mills lambda	–0.75	–6.25	–0.65	–5.0	–0.66	–5.07
Censored obs.	397		397		397	
Uncensored obs.	376		304		304	
Obs.	773		701		701	
Wald $\chi^2(1)$	17.61		29.16		35.47	
Prob>χ^2	0.000		0.0037		0.005	

Notes: (a) The regressions also include a constant term. ** p values < 0.05, * p values < 0.1.

Table 3.6 The impact of innovation strategies on firms' productivity growth: estimates

	Heckit[(a)]		*Heckit*[(a)]		*Heckit*[(a)]	
Estimation method	*Coefficient*	*t values*	*Coefficient*	*t values*	*Coefficient*	*t values*
Selection equation	*[1]*		*[2]*		*[3]*	
North-West	0.266*	1.87	0.358**	2.51	0.384**	2.52
North-East	0.184	1.26	0.232	1.57	0.247	1.57
Centre	0.184	1.23	0.173	1.14	0.234	1.43
South	Ref.	Ref.	Ref.	Ref.	Ref.	Ref.
D20_99	Ref.	Ref.	Ref.	Ref.	Ref.	Ref.
D100_249	0.421**	3.59	0.390**	3.50	0.463**	3.95
D250	0.578**	4.80	0.560**	4.85	0.663**	5.87
DTrade	0.437**	1.97	0.488**	2.15	0.468**	2.10
DHotel	0.008	0.03	0.204	0.80	0.170	0.69
DTrasp	0.165	0.78	0.271	1.22	0.246	1.13
DPost	6.460**	9.04	8.823**	7.61	7.573**	8.16
DFinan	4.133**	4.83	4.700**	4.18	4.362**	4.14
DComp	1.443**	3.60	1.626**	4.74	1.599**	4.46
DRDcon	0.924**	3.29	1.151**	4.26	1.114**	4.23
DLegmkt	0.889**	2.46	0.972**	3.02	0.934**	2.87
DOthbus	–0.285	–1.25	–0.156	–0.65	–0.188	–0.81
DWaste	Ref.	Ref.	Ref.	Ref.	Ref.	Ref.
DGroup	0.117	0.86	0.071	0.57	0.053	0.45
Second stage equation						
Ln(productivity$_{95}$)	–0.102**	–3.90	–0.091**	–3.41	–0.102**	–3.47

D_strategy_1	…	…	–0.068	–0.70	–0.062	–0.63
D_strategy_2	…	…	0.057	0.57	0.066	0.67
D_strategy_3	…	…	–0.027	–0.39	–0.030	–0.44
D_strategy_4	…	…	0.038	0.50	0.037	0.50
D_strategy_5	…	…	0.024	0.24	0.008	0.08
D_strategy_6	…	…	–0.003	–0.05	–0.022	–0.37
D_strategy_7	…	…	–0.074	–0.74	–0.066	–0.65
D_strategy_8	…	…	–0.108	–0.98	–0.096	–0.87
D_strategy_9	…	…	0.098	0.86	0.133	1.16
D_strategy_10	…	…	0.194**	2.49	0.203**	2.65
D_strategy_11	…	…	0.034	0.51	0.038	0.59
Sector dummies	No		No		Yes	
Size dummies	No		No		Yes	
Mills lambda	–0.66	–6.60	–0.59	–5.36	–0.62	–6.20
Censored obs.	397		397		397	
Uncensored obs.	376		304		304	
Obs.	773		701		701	
Wald $\chi^2(1)$	15.24		21.06		23.41	
Prob>χ^2	0.0001		0.0495		0.136	

Notes: (a) The regressions also include a constant term. ** p values < 0.05, * p values < 0.1.

with both negative (stress related) and positive effects emerging (Askenazy and Caroli 2006; Bain 1997; Gullie 2005). This may be interpreted as a valued added enhancing effect strictly related to a higher workforce team and higher individual productivity, spurred by associated organizational and HRM practices and even by better job-related motivations (Bryson *et al.* 2005; Cassidy *et al.* 2005; Matteucci *et al.* 2005). Thus, the *win–win* innovation–labour conditions (innovations strategy are in effect aimed at enhancing labour conditions and job satisfaction, e.g. HPWP such as total quality management, team working, job rotation etc.) scenarios implemented by firms may be the reason for our result.

Productivity growth effects

As expected, specifications that define labour productivity (turnover/employment) confirm previous results. We summarize the primary elements in our reasoning (Table 3.6). First, there is confirmation that innovation strategies linked to improvements in workforce job conditions impact on productivity growth as well as turnover. It is a flag of innovations on the Italian service sector.

Secondly, the drivers we observed as being significant for employment growth (extending market shares has a positive effect), are not present here. This means that stimulating employment growth (leading to lower productivity *ceteris paribus*) does not significantly reduce firm performance, probably because it also impacts positively on turnover (and in effect, although not significant, the sign for strategy seven is positive). The same is true for market-oriented strategies that produce a negative effect on employment: they do not improve productivity, since they also impact negatively on turnover. Thus, overall, the only strategy that pays off is adopting or investing in innovation changes that have a labour-related content, in terms of higher job satisfaction. If HRM, training and innovation are adopted jointly by firms, it is likely that productivity will be enhanced by such complementarity elements that do not undermine or that partly compensate for *management by stress* effects (Coriat 1995; 2002), enabling potential productivity gains. More specifically, the relationship between eco-innovations and job quality elements, intermediated by the links to other innovations and HRM strategies, is an unexplored area.

Third, and most relevant to here, employment and turnover effects seem to compensate for one another resulting in an observed insignificant relationship between eco-innovation strategies and productivity. In fact, this derives from a negative turnover impact. Eco-innovations strategies do not show positive links with productivity, but do not appear to undermine growth, even in a relatively short-term/market immature market scenario. In a long-run scenario of industry equilibrium adjusting to environmental strategies, then it cannot be assumed that productivity would increase: it will depend on the relative strength of eventual employment and turnover rebound growth effects, on which we commented above.

Conclusions

We studied the role of environmentally motivated innovation within the various innovative dynamics in services firms, by analysing the various links between innovation strategies (cost reduction, market, technological, organizational) and economic performances.

We exploited a unique merged database of 773 Italian service firms with 20 or more employees from 1993–1995 CIS data on firm innovation strategic motivations and 1995–1998 original firm performances data on employment and turnover.

Our main finding shows a negative link between the 'environmental motivations' of innovation strategies and both employment and turnover, while, consequentially, the effect on labour productivity is not significant. The two effects seem to balance each other.

The effect on employment is partly in line with past evidence and is not unexpected, presenting some similarities with the general Italian situation in services at that time. It may derive from efficiency improvements (dematerialization processes) which also have an impact by reducing the (rate of growth of the) workforce. In addition and more in depth, it could be the results of the adoption of end-of-pipe and (short sighted, merely efficiency-oriented) process innovations that could lead to a net destruction of jobs in absence of product innovations and a general long-run value added oriented environmental strategy of the firm, that may on the one hand destroy low-skilled labour units, but with a more than compensatory creation of skills and competencies linked to environmental goals of the organization.

It would be helpful for future research to disentangle the effects on low- and high-skilled workers, as well as analysing different service sectors. It is plausible that there is a net effect deriving from the destruction of low-skilled employment and the creation of more high-skilled jobs, as a consequence of increased environmental awareness and strategy. As far as environmental issues are concerned, then, heterogeneity across services branches is high and might turn out relevant. This is scope for fruitful future research.

The effect on turnover is a negative impact of environmental innovation strategy, implying either a short-medium effect, possibly balanced in the long run by net benefits in terms of higher added value, or a real negative (average) impact that may be contingent on the observed period, during which environmental considerations were not at the heart of strategic management policies for the majority of firms.

Despite this, productivity-related effects (the core of performance indicators) are not significant. If on the one hand, mainstream-oriented hypotheses regarding eventual negative impacts are thus not confirmed, although, as we have pointed out, on the other hand Porter-like effects and virtuous circles between environmentally strategies and performances do not seem to appear in this case study. We remark again that in-depth studies on specific service sectors are needed in order to point out within services heterogeneity with regard the environmental-economic performances links. Larger datasets are needed for such objectives.

Appendix

Table A.3.1 Variables description

Variable	*Description*
Geographic dummies	
North West	Liguria, Lombardia, Piemonte, Valle d'Aosta
North East	Emilia-Romagna, Friuli Venezia-Giulia, Trentino Alto-Adige, Veneto
Centre	Abruzzo, Lazio, Marche, Molise, Toscana, Umbria
South	Basilicata, Calabria, Campania, Puglia, Sardegna, Sicilia
Size dummies	
D20_99	20–99 employees
D100_249	100–249 employees
D250	259 and more employees
Sector dummies	
DTrade	Trade
DHotel	Hotel & Restaurants
DTrasp	Transport
DPost	Post & Telecommunication
DFinan	Financial services
DComp	Software & related
DRDcon	R&D, Engineering, Technical Consultancy
DLegmkt	Legal & Marketing
DOthbus	Security, Cleaning, Other Business Services
DWaste	Waste Disposal
Organisation	
DGroup	1 if the firm belongs to a business group, 0 otherwise
Innovative strategies	
D_strategy_1	Modify the array of services/products supplied
D_strategy_2	Extend the array of services/products supplied
D_strategy_3	Develop services with lower environmental impact/output
D_strategy_4	Maintain current market share
D_strategy_5	Increase market share
D_strategy_6	Penetrate new markets
D_strategy_7	Improve production/plant flexibility
D_strategy_8	Reduce production costs
D_strategy_9	Increase the quality of services/products
D_strategy_10	Improve the workforce job related welfare
D_strategy_11	Adapt technology to currently prevailing ones

Notes

1 We also estimated regressions assigning the value one to choices from (c) to (f). Some results differ, but not regarding the strategy [3] *develop services with lower environmental impact/output.*

2 Here we cannot assess the role of policies as the driver of innovation, or consequently performance. Nevertheless, if we exclude anticipation strategies, the period under observation is one when major policies were still not implemented at EU and national levels. We can assume therefore that such innovation strategies depend on firms' strategic management.

3 Even at the private profit maximization level, environmental issues are not excluded *a priori* by firms, but it could be that they are more integrated than expected within business strategies. The nexus between labour productivity and environmental productivity depends strongly on the existing interconnections at the technological level and at the level of the specific externalities addressed. It concerns the manifold employment, value added and environmental impact of environmental and non-environmental technology.

References

Askenazy, P. and Caroli, E. (2006) 'Innovative work practices, information technologies and working conditions: evidence for France', IZA Discussion paper series 2321, Bonn: IZA.

Bain, P. (1997) 'Human resource malpractice: the deregulation of health and safety at work in the US and Britain', *Industrial Relations Journal*, 28: 176–91.

Bryson, A., Forth, J. and Kirby, S. (2005) 'High involvement management practices, trade union representation and workplace performance in Britain', *Scottish Journal of Political Economy*, 52: 451–91.

Cainelli, G., Evangelista, R. and Savona, M. (2006) 'Innovation and economic performance in services: a firm level analysis', *Cambridge Journal of Economics*, 30: 435–58.

Calvo, J. (2006) 'Testing Gibrat's law for small young and innovating firms', *Small Business Economics*, 26: 117–23.

Cassidy, M., Gorg, H. and Strobl, E. (2005) 'Knowledge, accumulation, and productivity: evidence from plant level data for Ireland', *Scottish Journal of Political Economy*, 52: 344–58.

Cleff, T. and Rennings, K. (1999) 'Determinants of environmental product and process innovation – evidence from the Mannheim innovation panel and a follow-up telephone survey', *European Environment*, 5: 191–201.

Coriat, B. (1995), 'Incentives, bargaining and trust: alternative scenarios for the future of work', *International Contribution to Labour Studies*, 5: 131–51.

Coriat, B. (2002) 'Employee Participation and Organisational Change in European Firms. Evidence from a Comparative Overview of Ten EU Countries', CEPN-HDE, CNRS Research Units 7115, University Paris XIII, Paris: University Paris XIII.

Crespi, F. and Pianta, M. (2008) 'Diversity in innovation and productivity in Europe', *Journal of Evolutionary Economics*, 18: 529–45.

Evangelista, R. and Savona, M. (2002) 'The Impact of innovation on employment in services: evidence from Italy', *International Review of Applied Economics*, 16: 309–18.

Evangelista, R. and Savona, M. (2003) 'Innovation, employment and skills in services. Firm and sectoral evidence', *Structural Change and Economic Dynamics*, 14: 449–74.

Evans, D.S. (1987a) 'The relationship between firm growth, size and age: estimates for 100 manufacturing industries', *Journal of Industrial Economics*, 35: 567–81.

Evans, D.S. (1987b) 'Tests of alternative theories of firm growth', *Journal of Political Economy*, 95: 657–74.

Femia, A. and Panfili, P. (2005) 'Analytical applications of the NAMEA', Paper presented at the 2005 annual meeting of the Italian Statistics Society, Rome: Italian Statistics Society.

Gullie, D. (2005) 'Work pressures in Europe: trends and determinants', *British Journal of Industrial Relations,* 43: 351–75.

Kander, A. (2005) 'Baumol's disease and dematerialization of the economy', *Ecological Economics*, 55: 119–30.

Lotti, F., Santarelli, E. and Vivarelli, M. (2001) 'The relationship between size and growth: the case of Italian newborn firms', *Applied Economics Letters*, 8: 451–4.

Lotti, F., Santarelli, E. and Vivarelli, M. (2007) 'Defending Gibrat's law as a long run regularity', IZA Discussion Paper 2744, Bonn: IZA.

Mansfield, K. (1962) 'Entry, Gibrat's law, innovation and the growth of firms', *American Economic Review*, 52: 1023–51.

Matteucci, N., O'Mahony, M., Robinson, C. and Zwick, T. (2005) 'Productivity, workplace performance and ICT: industry and firm level evidence for Europe and the US', *Scottish Journal of Political Economy*, 52: 359–85.

Nelson, R. and Winter, S. (1982) *An Evolutionary Theory of Economic Change*, Belknap Press/Harvard, Cambridge MA: Cambridge University Press.

Pfeiffer, F. and Rennings, K. (2001) 'Employment impacts of cleaner production: Evidence from a German study using case studies and surveys, *Business Strategy and the Environment* 10(3): 161–175.

Rennings, K., Ziegler, A. and Zwick, T. (2001) 'Employment changes in environmentally innovative firms', Discussion paper 46, Centre for European Economic Research (ZEW), Mannheim: ZEW.

Rennings, K., Ziegler, A., Ankele, K., Hoffmann, E. and Nill, J. (2003) 'The influence of the EU environmental management and auditing schemes on environmental innovations and competitiveness in Germany. An analysis on the basis of case studies and a large scale survey', Discussion paper 03-14, ZEW, Mannheim.

4 An early assessment of the influence on eco-innovation of the EU Emissions Trading Scheme

Evidence from the Italian paper industry[1]

Serena Pontoglio

Introduction

The link between environmental policies and innovation has been extensively discussed in the theoretical literature and has been the object of several empirical studies.[2] The relevance of this issue is increasing with the emergence of global environmental problems that require technological solutions, and the development of progressively more demanding and ambitious regulation in the field of climate and energy.

Emissions trading refers to a group of market-based instruments (MBIs) or economic instruments (EIs) relating to pollution and its control. These regulatory instruments encourage responses to market signals rather than establishing explicit directives on firm behaviour (Stavins 2003). Emissions trading is based on the creation of a market that assigns a price to a determined environmental externality. If a scheme is able to rely on a good design, market price will act as an incentive to develop or adopt innovative solutions to reduce this externality even below the minimum legal level.

Despite extensive support for emissions trading systems (ETS), some of the theoretical literature that ranks environmental instruments according to their potential innovative spillovers, shows mixed results (Kemp and Pontoglio 2008; Vollebergh 2007; Taylor 2007; Requate 2005). Requate (2005), for instance, states that we do not have much empirical evidence on the extent to which market-based instruments perform, firstly because there has been little opportunity for empirical comparison of different instruments under similar conditions, and secondly, because many countries regulate the same externality using several different instruments. Criticism of emissions trading also emerges in Driesen (2003). Driesen asserts that the environmental economics literature is biased in presenting ETS as superior because the analysis is concentrated on the group of emitters who have an incentive to make extra reductions so that they can profit from the sale of credits, thereby showing extra reductions compared with equivalent policy based on traditional regulation. In Driesen's view, this excludes half of the actors involved, that is, those that comply with the existing regulation by buying permits on the market.

The application of ETS is confined mostly to US experiences.[3] Taylor (2007) provides a systematic analysis of the main US cap-and-trade systems in the field of air pollution. The policies analysed are the national sulphur dioxide (SO_2) trading scheme introduced by the 1990 Clean Air Act Amendments, also known as Title IV; the NO_x Budget Trading Program (NBP), a seasonal and regional scheme for nitrogen oxide emissions implemented in 1999, and RECLAIM, the Regional Clean Air Incentives Market which has been in force since 1994 in southern California and applies to NO_x and SO_2 emissions. Taylor demonstrates that under Title IV and NBP the main innovations related to emissions consist of end-of-pipe solutions (primarily post-combustion devices) or fuel switching. Under RECLAIM, due to an initial over-allocation of allowances, operators were not required to reduce emissions. Regulation of industrial SO_2 emissions in the US in the period 1995 to 1990 was based on command and control (CAC) instruments and market-based instruments (MBIs, such as the Clean Air Act). Taylor *et al.* (2005) and Taylor (2007) analyse patent data and show that the Clean Air Act has led to less innovative solutions compared with earlier regulation, and this effect is considered in strict connection with lower-than-expected allowance prices. This review reveals that the empirical literature is not sending a clear message that emissions trading experiences spur innovation. This is in part due to the limited number of studies in this area (OECD 2005), the limited scope of the experience reviewed, and contextual deficiencies in some of the systems analysed (Gagelmann and Frondel 2005). What also emerges from the empirical literature is that there are more aspects pertaining to policy setting and going beyond the nature of the policy instrument, that need to be investigated.

The EU ETS and innovation: an assessment of the elements influencing its innovative potential in Phase I

The EU ETS is the cornerstone of European climate policy and was introduced by the Directive 2003/87/EC. It covers some 10,800 industrial installations across Europe (EEA 2008), in four energy-intensive sectors: energy activities (power, heat and steam generators, oil refineries and coke ovens); production and processing of ferrous metals (iron and steel); minerals (cement, glass and ceramics); and pulp and paper. Its first application phase (Phase I) started January 2005 and covered the period 2005–2007; and the second phase overlaps with the first Kyoto commitment period of 2008–2012.

The design of environmental policies can have a major impact on their innovative potential. Based on the outcomes of the theory on tradable permits, in this chapter we identify design issues that enable the greatest influence of the EU ETS and their impacts on innovation.

It is important first to recall that the EU ETS is a decentralized scheme.[4] This reflects the EU institutional setting: Directive 2003/87/EC provides a framework of general rules and criteria, but delegates to member states the responsibility to establish the details of the design to be integrated with a National Allocation Plan (NAP). NAPs must be approved by the European Commission (EC) to ensure consistency with the common principles in Annex III of Directive 2003/87/EC.

The most important EU ETS design rules relate to: (a) the nature of the system; (b) coverage; (c) linkages; (d) allocation principles; (e) new entrant and closure rules; and (f) time profile.

In terms of the nature of the system, tradable permit schemes fall into two main categories: baseline-and-credit and cap-and-trade schemes. In the former, operators must respect a relative emissions technical threshold or benchmark, calculated for a homogeneous group of activities. The EU ETS belongs to this category and requires operators to respect a fixed limit (cap) on their emissions. According to the theory (Tietenberg 2001, 2003), provided that the two systems are equivalent, the incentives for innovation should be the same within either alternative. Thus, we consider this design element as being neutral with reference to innovations.

Coverage of the system refers to its size and the sectors or economic activities included in the scheme. The theory suggests that enlargement of the scheme (opt in) to new sectors or actors works in favour of its static efficiency, since new opportunities to reduce carbon emissions are made available. However, the wider opening of the scheme can have ambiguous effects on its potential innovative outcomes because, for example, the greater the number of economic activities involved, the higher the opportunities for innovation development or adoption. At the same time, the availability of carbon abatements at lower (higher) cost reduces (increases) the economic effort required, and the market price of carbon dioxide (CO_2). A reduction (increase) in the carbon price is reflected in a lower (higher) economic incentive to develop or adopt a carbon-saving innovation.

The situation is similar for linkages with other trading systems. Linkages are established according to the amended 'Linking Directive', which enables a connection between EU ETS and the Kyoto Protocol's project-based mechanisms. Emissions reduction credits generated in developing countries or countries with economies in transition, by means of the Clean Development Mechanism (CDM) and Joint Implementation (JI) may be used (up to a certain limit), as substitutes for ETS allowances.

Allocation methods and principles have both indirect and direct effects on the potential role of a tradable permit scheme in promoting eco-innovation. Their indirect effects relate to the impact of allocation decisions on the cap and its level of stringency. Direct impacts depend mainly on the single allocation rules. Many alternative methods apply to the initial distribution of permits in cap-and-trade systems. Permits may be distributed for free (grandfathered) or can be purchased by operators (i.e. auctioned). Economic theory asserts that the efficiency of the scheme is assured irrespective of initial permit allocation (Tietenberg 2001); however, a number of recent studies show that the issue is more controversial since allocation rules have distribution effects and can influence both static and dynamic efficiency (Vollebergh 2007). When permits are auctioned, operators are forced to make visible their marginal abatement costs and to declare their interest and ability to pay for an emissions reduction unit. Auctions ensure that there is no over-allocation of permits, since rational operators will not buy more permits than the amount necessary to cover their annual emissions.

In the EU ETS both allocation and cap-setting were decentralized: the EU ETS Directive did not impose an allocation method on national authorities, which, in Phase I, were free to set their own rules. This discretion has led to discrepancies in national plans, and made the need for harmonization of the EU ETS urgent in order to avoid distorting effects on competition. Despite this diversity, we can identify certain common basics in the allocation rules: the vast majority of the EU-27 countries distributed EUAs (EU emissions allowances) for free based on historical emissions, in many cases adjusting them according to projected growth. It has been shown that many of these projections were rather inflated (Egenhofer 2007). In some NAPs, allocations have been adjusted according to rules that reward the use of highly efficient technologies, or to compensate for early action. In other cases, allocations took account of efficiency parameters and reduction potential. The auctioning of allowances generally met with resistance and intense pressure from industry.

In relation to new entrants, the majority of member states opted to set aside some allowances for distribution, at no cost, to newcomers, on a first come, first served basis. Theory suggests that the most efficient rule for allocation to new entrants to stimulate innovation is purchase of allowances in the market or through auction. New entrants are fundamental to the diffusion of more efficient technologies and auctions increase the incentive to invest less in carbon-intense technologies. In terms of closures, the cancellation of allowances owned by operators that ceased their activities might seem logical, but does not allow full internalization of the opportunity costs of emissions. The withdrawal of allowances acts in fact as a disincentive to close inefficient plants and invest in new more efficient ones. EU ETS allowances allocated to plants that are subsequently closed have to be cancelled from the registry, although some member states have introduced a transfer rule for their transfer to new installations.

The timing of an environmental policy instrument is fundamental to a successful innovation-oriented policy. Temporal flexibility in a trading scheme is defined by banking and borrowing rules, which can have a negative impact on the promotion of innovation. Banking of permits allows for their use in subsequent compliance periods and, thus, rewards early emissions reductions, since current savings can be realized in the future. Borrowing introduces the possibility of using compliance permits assigned for future periods. The EU ETS allows for both banking and borrowing and the overall effect of these two mechanisms on innovation is ambiguous. Although there is no restriction on banking and borrowing, intra-period (within Phase I or Phase II), inter-period transfer of permits is disallowed. This limitation on temporal flexibility is based mainly on the necessity to keep EU ETS compliance separate from the Kyoto targets. However, as observed by Ellerman and Joskow (2008) this feature rendered Phase I self-contained and was a major cause of prices declining to zero in 2006–2007. Banking and borrowing are not the only design issues related to timing. The EU ETS consists of several temporal stages and its time profile is defined also by the length of its allocation periods, whose three to five years are far shorter than the investment cycles in energy-intensive industries. In addition, the practice of updating the base year in each allocation round undermines the incentive for an immediate emissions reduction.

Other elements that have an impact on the innovation decisions of operators are price volatility and status of international climate agreements. These are not EU ETS design issues, but are related respectively to market dynamics and international climate negotiations. Price volatility is a feature that is common to new tradable permits schemes (Taylor *et al.* 2005; Ellerman *et al.* 2003), but introduces further uncertainty. The collapse of the carbon price in May 2006 reduced market reliability for operators. In the international policy context, the absence of a post-2012 climate regime affects predictability and discourages long-term investments.

Based on this brief analysis of the main design issues that may impact on the innovative potential of the EU ETS, we can say that, for the most part, the indications in the literature on the best alternatives in terms of maximizing the power of emissions trading to promote innovation and incentivize the diffusion of low-carbon solutions, are not reflected in the design of the EU scheme. The EU ETS is far from an ideal mechanism, or at least was during its first period of application. Adjustments introduced in Phase II (2008–2012), such as NAPs harmonization, and the amendments proposed for Phase III (European Commission 2008), have improved its configuration considerably.

The response of the Italian paper industry to EU ETS

This section presents the results of a survey of the Italian paper industry, which investigated its position in the EU ETS (shortage/excess of allowances), compliance strategies and adoption of innovative technological solutions to reduce carbon emissions from paper mills.

In 2006, the Italian paper industry comprised 181 paper mills and 23,000 employees, and represented 10.1 per cent of European paper and cardboard products production. Paper manufacturing processes are very energy-intensive and absorb large amounts of electricity and steam. Carbon emissions from the paper industry are mostly proportionate to its energy consumption and fuel carbon-intensity, with negligible amounts of emissions from the paper manufacturing process. The strategies open to paper mills to reduce their carbon intensity include fuel switching, increasing energy efficiency (in both energy production and the manufacturing process), greater use of renewable energy, shifting to less energy-intensive products.

When the first verified data on emissions were released by the EC in May 2006 they revealed an excess of allocations over verified emissions equal to 44 Mtons of CO_2 for 2005. This surplus was 5 per cent of the total annual allowances and was deemed to be an over-allocation (Kettner *et al.* 2007). After the release of data on real emissions, the carbon price collapsed as a consequence of the adjustment between market operators' expectations and reported data. The aggregate position of Italy is different from the European average since, together with the UK, Spain, Ireland and Austria, Italy was a net buyer of emissions in EU ETS Phase I. In

2005 verified emissions amounted to 225,931,351 tons of CO_2. The deficit in allowances, representing net demand for Italy, was 9,781,110 tons of CO_2 (see Table 4.1).

The Italian National Allocation Plan for Phase I (2005–2007) awarded the paper industry an average of 5.09 Mt of CO_2 per year including the New Entrants Reserve (0.58 Mt of CO_2). The allowances allocated to paper mills represented 2.3 per cent of the national emissions cap. In the allocations for Phase II (2008–2012) the allowances distributed have remained at the amount allocated for 2007. In 2005 the number of paper plants included in the NAP was 163. Paper plants differ in size, and we calculate that 55 per cent of the emissions allocation in Phase I was produced by 25 large mills, each emitting over 50,000 tons of CO_2 per year. The 57 smallest installations were responsible for 5.75 per cent of the emissions allocation for Phase I.

In 2005–2007, the European paper industry received excess allowances compared with produced and verified emissions. However, the Italian paper industry is itself a net-buyer and is an exception in the aggregate picture for Europe. To simplify, we can compare the case of two sample paper plants, one representing the Italian paper industry (plant A) and the other representing the paper industries in the rest of Europe (plant B). While plant A in 2005 and 2006 received allowances 4 per cent lower than its annual emissions, plant B received emissions allowances 20 per cent higher than its effective emissions (see Table 4.2).

The emissions shortfall in 2005 for the Italian paper industry amounted to 197,103 tons of CO_2. In 2006 this deficit was reduced to 116,945 tons of CO_2 or 2.4 per cent of the total allocations. The economic impact of the emissions shortage for 2005 was 0.5 per cent of the industry turnover (at a carbon price of €20/ton of CO_2). However, if we consider that only a small fraction of the shortage registered in 2005 was covered by the purchase of allowances on the market and the real

Table 4.1 Difference between allowances and verified emissions for Italy (tCO_2)

Year	*Allocation tCO_2 (2005)*	*Verified emissions tCO_2*	*Shortage tCO_2*	*Shortage (%)*
2005	216,150,241	225,931,351	–9,781,110	4.5%
2006	216,150,241	225,208,364	–9,058,123	4.2%

Source: Elaboration CITL data (Community Independent Transaction Log).

Table 4.2 Allowances and verified emissions

Year	*EUAs tCO_2*	*Verified emissions tCO_2*	*Shortage tCO_2*	*Shortage (%)*
2005	4,841,897	4,841,897	–197,103	4.0%
2006	4,841,897	4,841,897	–116,945	2.4%

Source: Own calculation, Community Independent Transaction Log, 2008.

market price of carbon in 2006/2007 was very low, this percentage can be reduced twenty-fold, resulting in a very modest economic impact.

Aggregate data on national emissions per sector provide a first overview of the position of the Italian paper mills in the EU carbon market. To analyse in more detail the response at company level to the imposition of a marketable limit on carbon emissions, and to look for micro-evidence of innovation strategies, we administered a survey questionnaire. The survey was mailed in May/June 2006 to all paper plant operators included in the NAP, and was accompanied by a text describing the purpose of the study and assuring anonymity to respondents. Of the 163 operators contacted, 38 responded, a response rate of 23 per cent. The survey responses were supported by interviews with industry experts (carbon consultants, engineers, and officials of the Italian Paper Industry Federation – Assocarta).

The responses show that, on average, 65 per cent of respondents were short of emissions allocations in 2005, 2006 and 2007. This figure is consistent with the overall performance in the sector. Seven of the respondent installations had requested an extra allocation of allowances based on substantial changes to their equipment (increased capacity).

In the case of shortfalls in allowances, operators have various options. They can buy allowances directly from another operator (bilateral exchange) or they can purchase them on the market, through the intermediation of a consultant or a dedicated platform. Where appropriate, operators can receive allowances from other installations in the same company group (intra-firm exchange). Finally, EU ETS allows for quota borrowing from allocations in subsequent periods although this was not a possibility for 2007 since it was not possible to make intra-period borrowing (see Figure 4.1).

The most common strategy adopted for 2005 and 2006 respectively, was borrowing of allowances, taken up by 72 per cent and 64 per cent of respondents short of allowances. For 2007, 62 per cent of the respondents acknowledged that they would have to take recourse to purchasing allowances on the market to cover their cumulative shortage.

In cases of excess allowances, operators can sell them on the market, exchange them with other firms in the same group, or accumulate them for use in a subsequent year (banking). Respondents with excess allocation in 2005 and 2006 mainly took the banking option (83 per cent).

Thus, most respondents adopted a 'wait and see' strategy characterized by conservative and cautious decision making and high-level use of time-flexible solutions. Years 2005 and 2006 were characterized by uncertainty about EU ETS administrative management and the economic consequences; this was due mainly to the novelty of the instrument and to the delays that accompanied the process of implementation of the EU ETS Directive in Italy. In such a context, caution prevails over proactive strategies and many operators decided to postpone abatement decisions to later years.

The survey also asked about collateral activities beyond compliance with the EU ETS. More than 30 per cent of respondents used consultants to manage the administration related to EU ETS compliance rules; 30 per cent of respondents

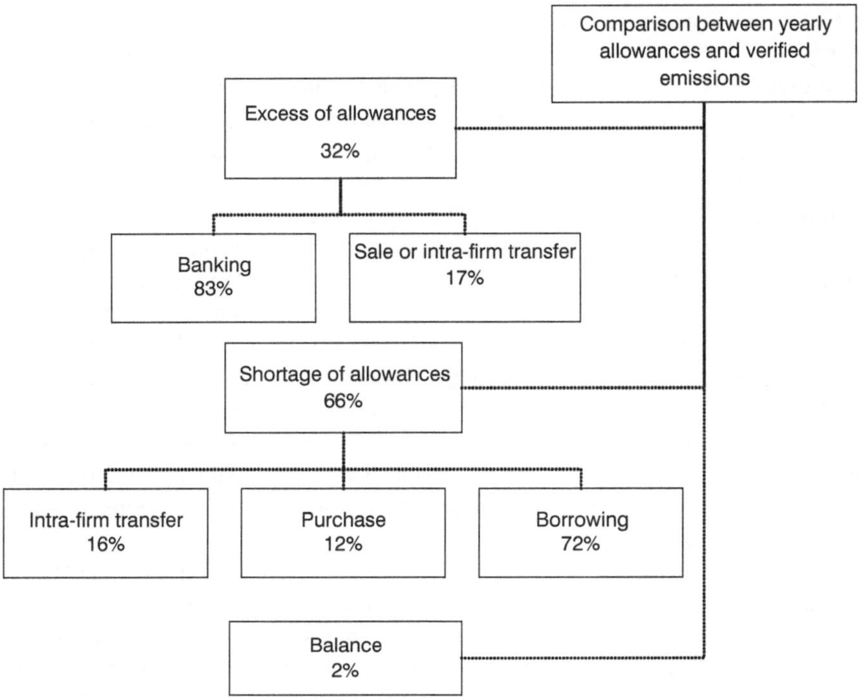

Figure 4.1 Respondents' compliance strategy (%)

engaged in training to learn about the ETS mechanisms. These results show that firms allocated resources in Phase I to learn about emissions trading and the accompanying administrative and technical requirements (CO_2 monitoring, communication and verification procedures). None of the respondents were involved in CDM/JI or Carbon Funds. A section of the questionnaire contained questions about the effect of the EU ETS on investment strategies and innovation decisions. A majority of respondents (52 per cent) declared no actions or plans to introduce technological innovations aimed at reducing CO_2 emissions, but 13 per cent responded that they had undertaken innovation and 35 per cent were developing projects to achieve changes in subsequent years. A simple correlation analysis demonstrates that firm size has no influence on decisions related to innovative technological solutions. In terms of the typology of innovations adopted or planned, 38 per cent were energy-efficient solutions and 23.8 per cent were designed to optimize production. This latter category includes developments to reduce raw materials usage (water and fibre), which resulted in a reduction in energy inputs (and carbon emissions). Few operators had opted to switch fuel and none had plans to change their raw material inputs or sets of products offered.

The availability of new technologies to reduce carbon emissions below current levels is fundamental to the realization of carbon-saving interventions. We would refer here to the reduction potential of a plant, machinery, an individual installation or a sector. The reduction potential can be calculated as the abatements in emissions obtainable through the substitution of the current technology (new machinery, new system of production, etc.) to achieve lower carbon intensity. We conducted interviews with experts and machinery suppliers in order to learn about the potential for reductions in the Italian paper industry. Our interviewees agreed that Italian paper plants are old on average and for this reason there are significant margins for improving both energy efficiency and use of raw materials, which would both help to reduce CO_2 emissions. It is not possible to identify a single technology that would result in an outstanding reduction in carbon emissions, but there is a variety of solutions that could be adopted, along the production process, each of which would yield a specific improvement.

Conclusions

The extent to which the EU ETS creates a context favourable to the development and diffusion of innovations is a key criterion against which EU ETS should be assessed. Technology advancements are crucial for the attainment of the ambitious European level target of a reduction of 20 per cent of carbon emissions by 2020.

The analysis of EU ETS design during Phase I provides a picture of a system that does not encourage innovation and does not reward enough for investment in low-carbon solutions. If environmental policies in theory promote the adoption and diffusion of more carbon-friendly technological solutions, the flawed design of EU ETS Phase I reduces this potential. In particular, there are some major differences between the EU ETS framework and an ideal emissions trading mechanism in terms of allocation principles, rules for new entrants and closures and time profile issues.

The national cap-setting process has led to insufficiently strict targets in Phase I and in a decline in prices towards zero. Emissions trading is an economic instrument and the presence of clear and visible economic signals is vital for its proper working. In the presence of weak or zero prices for carbon reductions it is not possible to realize the economic gains associated with the realization of investments in low-carbon solutions.

Our investigation of the reactions of operators in the Italian paper industry show that they have adopted a wait-and-see strategy, characterized by conservative and cautious decision making and use of time-flexibility solutions. These are having modest outcomes in terms of innovation. Carbon dioxide emitted by energy-intensive industries cannot be reduced through the use of low-cost end-of-pipe abatement solutions; they require improvements in energy-efficiency and investment in renewable energy. The adoption of these technologies is influenced by both energy and carbon prices and the actors involved in their development and diffusion are machinery suppliers who are the main players in the paper innovation system.

EU ETS Phase II encompasses some changes, and significant improvements are been proposed for the post-2012 EU ETS. It will be interesting to see whether these changes account for the flaws in the current EU ETS, and create incentives for clean investments. There are some positive signs, such as the caps approved for Phase II, which, after the conclusion of the Commission's assessment appear to be some 13 per cent lower than in the first period cap and 6 per cent lower than for 2005 emissions levels.

Notes

1 The views in this chapter are those of the author and do not necessarily represent the position of her institution.
2 In this chapter we adopt a broad definition of innovation in low-carbon technological solutions, which refers to the production, assimilation or exploitation of a product, production process, service or management or business method that is novel to the firm and which results, throughout its life cycle, in reductions in carbon-intensity of the installations into which it is introduced.
3 In Europe, before EU ETS came into force, national emissions trading schemes had been implemented in the UK and Denmark to regulate carbon emissions in the power generation sector. To these experiments we should add intra-firm schemes implemented by British Petroleum and Shell in the 1990s.
4 In this respect, Ellerman and Joskow (2008) observe that the decentralized nature of the cap is the main difference between the EU ETS and earlier US cap-and-trade experiments.

References

Driesen, D.M. (2003) 'Does emissions trading encourage innovation?', *Environmental Law Reporter,* 1: 10094–10108.

Egenhofer, C. (2007) 'The making of the EU Emissions Trading Scheme: Status, prospects and implications for business', *European Management Journal*, 25: 411–14.

Ellerman, D. and Joskow, P.L. (2008) 'The European Union's Emissions Trading System in perspective', Arlington, VA: Pew Center on Global Climate Change.

Ellerman, D., Joskow, P.L. and Harrison, D. (2003) 'Emissions trading in the US: Experience, lessons and considerations for greenhouse gases', Arlington, VA: Pew Center on Global Climate Change.

European Commission (2008), *Proposal for a Directive of the European parliament and of the Council amending Directive 2003/87/EC so as to improve and extend the greenhouse gas emission allowance trading system of the Community*, COM(2008) 16, Brussels: European Commission

EEA (European Environmental Agency) (2008), *Application of the Emissions Trading Directive by EU Member States – Reporting Year 2007*, EEA Technical report 3/2008, Copenhagen: EEA.

Gagelmann, F. and Frondel, M. (2005) 'The impact of emission trading on innovation – science fiction or reality?', *European Environment*, 15: 203–11.

Kemp, R. and Pontoglio, S. (2008) 'The innovation effects of environmental policy instruments. A typical case of the blind men and the elephant', Paper for DIME WP 2.5 Workshop on Empirical Analyses of Environmental Innovations, Fraunhofer Institute for Systems and Innovation Research (ISI), Karlsruhe, 17–18 January, 2008 Karlsruhe: ISI.

Kettner, C., Köppl, A., Schleicher, S.P. and Therius, G. (2007) 'Stringency and distribution in the EU Emissions Trading Scheme – the 2005 evidence', FEEM Working paper 22, Milan: FEEM.

OECD (2005) *New Commitment Options: Compatibility With Emissions Trading*, Paris: OECD

Requate, T. (2005) 'Dynamic incentives by environmental policy instruments – A survey', *Ecological Economics* 54 (2–3), p. 175–195.

Stavins, R.N. (2003) 'Experience with market-based environmental policy instruments', in K. Maler and J. Vincent (eds), *The Handbook of Environmental Economics,* Amsterdam: North-Holland,, 1, p. 355–435

Taylor, M.R. (2007) 'The Dynamics of Innovation and Cap-and-Trade Programs', *Working Paper*, California: University of Berkeley.

Taylor, M.R., Rubin, E.S. and Hounshell, D.A. (2005) 'Regulation as the mother of innovation: The case of SO_2 control', *Law & Policy*, 27: 348–78.

Tietenberg, T. (2001) 'Emissions Trading Programs', vol.I and II, International Library of Environmental Economics and Policy.

Tietenberg, T. (2003) 'The tradable permits approach to protecting the commons: lessons from climate change', *Oxford Review of Economic Policy*, 19: 3.

Vollebergh, H. (2007) 'Impacts of Environmental Policy Onstruments on Technological Change', OECD Report, 7 February 2007, Paris: OECD.

Part II

Environmental efficiency and sector performances

5 Emissions trends and labour productivity dynamics

Giovanni Marin and Massimiliano Mazzanti

Introduction

Indicators of delinking or decoupling, that is, improvements of environmental/resource indicators with respect to economic indicators, are increasingly used to evaluate progress in the use of natural and environmental resources. Delinking trends for industrial materials and energy in advanced countries have been under scrutiny for decades. In the 1990s, research on delinking was extended to air pollutants and greenhouse gases (GHGs, henceforth) emissions. 'Stylised facts' were proposed on the relationship between pollution and economic growth that came to be known as the Environmental Kuznets Curve (EKC, henceforth), which was based on general reasoning around relative or absolute delinking in income–environment dynamics relationships.

The value of this mainly empirical analysis is manifold. First, its originality lies in the very rich NAMEA sector-based economic-environmental merged dataset for 1990–2005 (29 branches), which is further merged with data on trade openness for the EU_{15} and extra-EU_{15} dimensions, and research and development (R&D) sector data. The quite long dynamics and the high sector heterogeneity of these data allow robust inference on various hypotheses related to the driving forces of delinking trends. In this analysis, we investigate CO_2 and SO_x. In addition to core evidence on the EKC shape, we test the following hypotheses: (a) whether services and industry have moved along different directions; (b) whether the increasing trends associated with trade openness among the EU_{15} and non-EU_{15} countries affect emissions dynamics, following the pollution haven debate (Cole 2003, 2004; Cole and Elliott 2003; Copeland and Taylor 2004); (c) whether pre-Kyoto and post-Kyoto dynamics show different empirical structures; (d) whether sector R&D plays a role in explaining emissions efficiency. As empirical reference models we use a standard EKC model that measures emissions in relation to employees as an indication of environmental technical efficiency, and a STIRPAT/IPAT model, which uses emissions as the dependent variable, and relaxes the assumptions about unitary elasticity with respect to labour (population), which enters as a driver. The policy relevance of this work lies in: (i) the temporal structural break in pre-Kyoto and post-Kyoto dynamics; and (ii) the macro sector (services, industry) level evidence it provides which could help to shape EU policies such as refinements to existing Emission Trading Scheme (ETS), or a new carbon tax for non-industry sectors or small businesses. The use of NAMEA accounting, which is a panel (that consists of a time series of cross-sections) of observations for air pollutants, value added, trade, R&D and employment matched for the same productive branches of the economy (Femia and Panfili 2005), is a novelty of our study, compared with other international studies on EKC. We use a disaggregation at 29 productive branches and four air polluting emissions.

For brevity, we refer the reader to Marin and Mazzanti (2009) and Mazzanti and Zoboli (2005, 2009) for a conceptual discussion of IPAT and EKC frameworks.

Empirical model and data sources

Models and research hypotheses: EKC oriented specifications

We test two kinds of models: the first uses the EKC framework as a reference (Mazzanti *et al.* 2008a,b); the second is a modified STIRPAT model.[1]

We reformulate the EKC relationship to exploit the sector-level disaggregation of the NAMEA matrix. This framework means we lose standard demographic and income information, but allows us to take advantage of insights on economic and environmental efficiencies in the production process. Equation (5.1) shows the EKC based empirical model:

$$\ln(E_{it}/L_{it}) = \beta_{0,i} + \beta_1 Kyoto_{0,1} + \beta_2 \ln(VA_{it}/L_{it}) + \beta_3[\ln(VA_{it}/L_{it})]^2 + \beta_4[\ln(VA_{it}/L_{it})]^3 + \varepsilon_{it} \qquad (5.1)$$

In Equation (5.1) *environmental technical efficiency* (emissions/full-time equivalent jobs) is a function of a third order polynomial equation of labour productivity (in terms of value added per full-time equivalent job), individual (sector) dummy variables (β_{0i}) and a temporal structural break called *Kyoto*, coded zero for 1990–1997 and one for 1998–2005. The logarithmic form of the dependent and explanatory variables enables the estimated coefficients to be interpreted as elasticity. We test Equation (5.1) on the whole dataset (29 branches) and then on the separate industry (C, D, E and F) and services (G to O) macro-sectors in order to check whether the average picture differs from that provided by the sub-sample results.

Third order polynomial form allows us to test for non-linearity (normal/inverted U or N-shaped curves) in the relationship between E/L and VA/L. A significant cubic specification results in N (or inverted N) shaped curves, while a quadratic one signals U (or inverted U) curves. The choice of polynomial order in the EKC literature is still somewhat controversial. First, an N-shaped curve indicates that absolute delinking is followed by a return to a monotonic joint trend of environmental pressures and economic growth (re-coupling) determined by a strong scale effect. Second, many authors (Stern, 1998) point out that both forms allow environmental pressure tending to an infinite plus or minus, both physically impossible outcomes. Finally, an N (or inverted N) shaped curve in a medium-short period perspective may indicate a rather volatile relationship. We believe it is relevant to assess these non-linear shapes in our framework, given that we analyse dynamic relationships across different sectors and pollutants. In addition, even in the presence of pollutants already showing evidence of absolute delinking, the re-coupling hypothesis is worth investigating as a possible (new) state of the world.

Individual effects (β_{0i}) capture the specific features of the branch in terms of average emissions intensity. We estimate these individual effects using a fixed effects model (FEM) following Wooldridge (2005).

In addition to the core specification, we design a sort of Kyoto structural break by means of a dummy variable that takes the value one for the years after 1997, to try to capture the direct or indirect effects of the Kyoto Protocol: the direct effects should be GHG emissions reductions in response to policies introduced to meet the Kyoto target; the indirect effects will be related to the anticipatory strategies for future policies on GHGs and, for pollutants, from the ancillary benefits from GHG emissions reductions.[2] We can state, therefore, in addition to specific Kyoto-related effects, this dummy variable captures temporal variations in emissions linked to various policy effects in the EU and Italian environment, and other temporal changes common to all the branches. The antilog of β_1 can be viewed as the average level of emissions *ceteris paribus* in 1998–2005, with average emissions levels in 1990–1997 equal to one.

We first extend the base model by adding two *trade openness indexes*, one for the EU_{15} and one for the extra-EU_{15} area. Because of the high level of correlation between the two openness indexes (0.796) we analyse them separately to overcome potential collinearity problems. We can then refer to Equations (5.2) and (5.3):

$$\ln(E_{it}/L_{it}) = \beta_{0i} + \beta_1 Kyoto_{0,1} + \beta_2 \ln(VA_{it}/L_{it}) + \beta_3[\ln(VA_{it}/L_{it})]^2 + \beta_4[\ln(VA_{it}/L_{it})]^3 + \beta_5 \ln(TO_{EU15})_{it} + \varepsilon_{it} \quad (5.2)$$

$$\ln(E_{it}/L_{it}) = \beta_{0i} + \beta_1 Kyoto_{0,1} + \beta_2 \ln(VA_{it}/L_{it}) + \beta_3[\ln(VA_{it}/L_{it})]^2 + \beta_4[\ln(VA_{it}/L_{it})]^3 + \beta_5 \ln(TO_{EXTRA\_EU15})_{it} + \varepsilon_{it} \quad (5.3)$$

For a review of the theoretical reasoning behind the link between trade openness and emissions growth, we refer among others to Zugravu *et al.* (2008), Frankel and Rose (2005), Cole (2003, 2004, 2005), Cole and Elliott (2003), Dietzenbacher and Mukhopadhay (2007) and Mazzanti *et al.* (2008a,b). The sign of the relationship depends on two potentially conflicting forces: the delocalisation of polluting industries in less-developed areas with lax regulation (*pollution haven* effect); and the country specialisation in capital intensive and energy intensive industrial sectors (*factor endowment* effect). The originality of our empirical exercise is that we are able to disentangle two trade openness dynamics, within EU_{15} and extra-EU_{15}. We can state here that EU_{15} openness is not expected to be associated to *pollution haven* effects on the basis of the growing homogeneity of European environmental policies: we can expect then either a not significant or a negative effect on emissions. EU environmental policies explicitly take account of and correct for potential intra-EU unwanted and harmful to the environment displacement of polluting productions in search of lax environmental policies. Such homogeneity, linked to the growing stringency in EU-wide environmental regulations, could result in a high correlation between EU_{15} openness and the stringency of domestic environmental regulation, with a potential beneficial effect (*race-to-the-top*) on environmental efficiency. In the contingent case of Italy, the main trade relationship with Germany, a leader in (environmental) technology and standards in the EU, is a relevant anecdotal fact. Communitarian openness, apart from race-to-the-top effects, is related to intra-sector specialisation in response to relative abundance/scarcity of factors (linked to particular environmental pressures) endowment and the spread of environmental efficient technologies.

Extra-EU_{15} openness instead captures the balance between the *factor endowment* and *pollution haven* effects: Italy is expected to have a comparative advantage in capital (and then pollution) intensive productions and more stringent environmental regulation relative to the average extra-EU_{15} trade partners; even relying on the empirical evidence on the issue of environmental effects of trade openness, we can state that no *a priori* expectation about the sign of the relationship between extra-EU_{15} openness and environmental efficiency is possible.

Finally, we test the effect of R&D/VA, in order to evaluate whether the innovative efforts of enterprises could have a beneficial or negative effect on environmental efficiency. Generally, the adoption of process/product innovations occurs with a delay as a consequence of R&D investments. We use a contemporary R&D/VA ratio because if we use lags we lose too many observations. If we add R&D, Equation (5.4) becomes the estimation basis.

$$\ln(E_{it}/L_{it}) = \beta_{0i} + \beta_1 Kyoto_{0,1} + \beta_2 \ln(VA_{it}/L_{it}) + \beta_3[\ln(VA_{it}/L_{it})]^2 + \beta_4[\ln(VA_{it}/L_{it})]^3 + \beta_5 \ln(R\&D_{it}/VA\_curr_{it}) + \varepsilon_{it} \quad (5.4)$$

Models and research hypotheses: STIRPAT based specifications

The second category of models is an adaptation to a single-country sector disaggregation of the STIRPAT framework (Dietz and Rosa 1994; York *et al.* 2003). The stochastic reformulation of the IPAT formula relaxes the constraint of unitary elasticity between emissions and population, implicit in EKC studies where the dependent variable is the logarithm of per capita pressures on the environment (Martinez-Zarzoso *et al.* 2007; Cole and Neumayer 2004). This model allows us to investigate explicitly the role of demographic factors in determining environmental pressures and to use a non-relative measure of this pressure as the dependent variable.

We start from a revised IPAT identity, as described in Equations (5.5)–(5.8) below, where the emissions (E) for each branch are the multiplicative result of employment (L), labour productivity (VA/L) and emission intensity of value added (E/VA).

$$E = L*(VA/L)*(E/VA) \quad (5.5)$$

$$E_{it} = \delta_{0i} * (L_{it})^{\delta_1} * (VA_{it} / L_{it})^{\delta_2} * (E_{it} / VA_{it})^{\delta_3} * e_{it} \tag{5.6}$$

$$\ln(E_{it}) = \delta_{0i} + \delta_1 \ln(L_{it}) + \delta_2 \ln(VA_{it} / L_{it}) + \delta_3 \ln(E_{it} / VA_{it}) + \ln(e_{it}) \tag{5.7}$$

$$\ln(E_{it}) = \beta_{0i} + \beta_1 \ln(L_{it}) + \beta_2 \ln(VA_{it} / L_{it}) + \varepsilon_{it} \qquad (5.8)^3$$

$$\begin{aligned}\ln(E_{it}) = \beta_{0i} + \beta_1 Kyoto_{0,1} + \beta_2 \ln(VA_{it} / L_{it}) + \beta_3 [\ln(VA_{it} / L_{it})]^2 \\ + \beta_4 [\ln(VA_{it} / L_{it})]^3 + + \beta_5 \ln(L_{it}) + \beta_6 \ln(L_{it})^2 + \varepsilon_{it}\end{aligned} \tag{5.9}$$

The above stochastic reformulation of Equation (5.5) has some interesting features: it allows separate investigation of the relationship between environmental pressures and employment and uses absolute pressures, which are related more to sustainability issues than relative ones, as the dependent variable. We should stress that in our analysis the focus is on labour not population. This opens the window to complex theory and empirical assessment of labour dynamics associated with technological development, and then with emissions dynamics. For the sake of brevity, we just touch on this issue referring the reader to other streams of the literature. To sum up, the relationship between emissions and employment recalls and is strictly connected to both the (dynamic) relationship between physical capital and labour and the relationship between emissions and physical capital. This relationship can identify particular effects associated with technological change: emission-saving effect, labour-saving effect and neutral effect.

We maintain the third order polynomial form for labour productivity and add the squared term of employment to test for non-linearities. Individual effects, the Kyoto structural break and labour productivity are interpreted similar to the EKC models, the difference being that they now refer to total, not per-employee, measures of environmental pressures, which may be more relevant for effective sustainability assessment and provided that policy targets are defined in total terms. The interpretation of the coefficients of employment varies depending on an increasing or decreasing level of labour. In the presence of increasing employment, we observe an emissions-saving effect when emissions increase less than proportionally to employment (or even decrease) (elasticity lower than one), whereas an increase more than proportional of emissions in comparison with employment shows a labour-saving effect (elasticity higher than one). When employment is decreasing the effect linked to each range of elasticity values is inverted.

Similar to the EKC equation, we test the STIRPAT based model on the whole dataset (29 branches) and on the separate industry and services macro-sectors. We add trade openness indexes and the *R&D/VA* ratio (equations not shown for brevity): the explanatory role of these variables in the model is the same as in the EKC framework.

The data

The contribution of our empirical analysis is as follows. Firstly, we assess EKC shapes for CO_2 and SO_x (evidence on more gases and emissions is provided in Marin and Mazzanti, 2009), using panel data disaggregated at sector level. We argue that using sector disaggregated panel data improves understanding of the income–environment relationship because it provides rich heterogeneity.

Secondly, we analyse the EKC shapes for total industry and services separately, in order to check whether the average picture differs from the sub-sample results. The sub-sample analysis is suggested by the conceptual perspective of NAMEA (Femia and Panfili, 2005). In the current work, we are specifically interested in exploring whether the income–environment EKC dynamics of the decreasing (in GDP share) manufacturing sector (more emissions-intensive), and the increasing (in GDP share) services sector (less emissions-intensive), differ. Additional drivers of emissions intensity are then included in order to control the robustness of main specifications and investigate further theoretical hypotheses. The main factors we investigate are trade openness, R&D and some policy-oriented proxies.

We use NAMEA tables for Italy for the period 1990–2005, allowing branch disaggregation at the 2-digit NACE (Ateco) classification level. NAMEA is produced by Istat, the national statistical agency. In the NAMEA tables environmental pressures (for Italian NAMEA air emissions and virgin material withdrawal) and economic data (output, value added, final consumption expenditures and full-time equivalent job) are assigned to the economic branches of resident units or to the household consumption categories directly responsible for environmental and economic phenomena. We use only data on economic branches, excluding household consumption expenditure and respective environmental pressures, with a disaggregation of 29 branches (2 for the agricultural sector, 18 for industry and 9 for the services sector). The added value of using environmental accounting data comes from the definitional internal coherence and consistency between economic and environmental modules and the possibility of extending the scope of analysis, but still maintaining this coherence and consistence.

We exploit the possibility of extending the basic NAMEA matrix by the addition of foreign trade data: for each branch, import and export of the items directly related to the output of the branch are included (CPAteco classification). Exports correspond to the part of the output of each NACE branch sold to non-resident units; imports are CPAteco domestically produced items bought by resident units (including households final and intermediate consumption) supplied by non-resident units. Data on national accounting for foreign trade are available from supply (import) and use (export) tables at the 2-digit CPAteco classification level (51 items) for the period 1995–2004. Istat also produces COEWEB, a very detailed database on Italian foreign trade: time series 1991–2005 of external trade are available at the level of the 4-digit CPAteco classification for A to E capital letters (agricultural sector and industry except F), with a disaggregation for the area (EU_{15}, EU_{25}, EU_{27} or extra-EU_{15}) of the partner. Unfortunately, we cannot exploit this database consistently because, for privacy protection reasons, Istat does not publish data for branches with less than three units: data related to such branches are also not included in the 4-digit disaggregation or the less detailed disaggregations. However, we use these distinctions between EU_{15} – extra-EU_{15} trade as a weight to split national accounting data. Hereafter, we construct trade openness indicators dividing the sum of imports and exports of every CPAteco category by the value added of the corresponding NACE branch:

$$(TO_{EU15})_{it} = \frac{(X_{EU15})_{it} + (M_{EU15})_{it}}{VA\_curr_{it}} \tag{5.10}$$

$$(TO_{EXTRA\_EU15})_{it} = \frac{(X_{EXTRA\_EU15})_{it} + (M_{EXTRA\_EU15})_{it}}{VA\_curr_{it}} \tag{5.11}$$

where X is export, M is import, *VA_curr* is value added at current prices, i is the branch (NACE) or the product (CPAteco) and t is the year between 1995 and 2004, the period of reference for the estimates using these covariates.

We also merge NAMEA tables with ANBERD OECD Database containing R&D expenditures of enterprises for 19 OECD countries, covering the period 1987–2003 (for Italy only 1991–2003, thus the period of reference in below regressions). Enterprises' expenditures are disaggregated according to the ISIC Rev. 3 standard, which is not perfectly compatible with NACE classification because it excludes units belonging to institutional sectors different from private enterprises. We retain only the industrial branches (excluding CA and CB) and exclude services. We use the R&D/VA ratio to derive information on the relative measure of innovative effort of the different branches and to get an index in constant prices. Because of the limited compatibility with national and environmental accounts, the ratio *per se* has a limited meaning but its variations may highlight changes in the relative innovative efforts of the enterprises in each industry branch.

For brevity we refer the reader to Mazzanti and Marin (2009) who present figures on the dynamics of aggregates and visual evidence on how outlier sectors affect and influence the non-linear shapes. We refer to Tables 5.1 to 5.3 for general overviews on NACE sectors and variables used for the analysis.

Empirical evidence

CO_2, EKC specifications

The evidence for CO_2 signals a relative delinking in the cases of the aggregate economy and industry: the elasticity of emissions efficiency with regard to labour productivity in fact is around 0.47–0.49. This outcome is as expected given that Italy is still lagging behind the Kyoto target. Table 5.4 presents the main regressions related to the comments in the text.

For services, estimates show a recoupling trend for CO_2, with a low turning point occurring within the range of observed values. This case highlights the

Table 5.1 Nace branches classification

	Nace (sub-section)	*Sector Description*
	A	Agriculture
	B	Fishery
Industry	CA	Extraction of energy minerals
	CB	Extraction of non energy minerals
	DA	Food and beverages
	DB	Textile
	DC	Leather textile
	DD	Wood
	DE	Paper and cardboard
	DF	Coke, oil refinery, nuclear disposal
	DG	Chemical
	DH	Plastic and rubber
	DI	Non metallurgic minerals
	DJ	Metallurgic
	DK	Machinery
	DL	Electronic and optical machinery
	DM	Transport vehicles production
	DN	Other manufacturing industries
	E	Energy production (electricity, water, gas)
	F	Construction
Services	G	Commerce
	H	Hotels and restaurants
	I	Transport
	J	Finance and insurance
	K	Other market services (real estate, ICT, R&D)
	L	Public administration
	M	Education
	N	Health
	O	Other public services

relevance of relying on and studying sector-based data. In fact, the recoupling vanishes, becoming an (expected) absolute delinking (negative linear relationship with elasticity –0.61) when we omit sector K (real estate, renting and business activities), a sort of outlier in this and other cases which we comment on below.

Focusing on the Kyoto structural break, the dummy presents a positive sign for CO_2 driven by industry dynamics; all the other estimated coefficients (for services and methane in industry) are negative. It seems, therefore, that neither the Kyoto emergence nor the 2003 Italian ratification has had significant effects on emissions performances by the main emitters, industrial sectors (76.28 per cent in 2005). Industry has not massively adapted to the new climate change policy scenario, and even the environmental Italian policy as a whole has somewhat lagged behind

Table 5.2 Correlation matrix

	ln(VA/L)	*ln(L)*	*ln(TO$_{EU15}$)*	*ln(TO$_{extraEU15}$)*	*ln(R&D/VA)*
ln(CO_2)	0.0887	0.0971	–0.1386	–0.1465	0.1036
ln(CH_4)	0.0134	–0.19	–0.2138	–0.2829	–0.0138
ln(NO_x)	0.2905	0.2157	–0.3233	–0.5185	0.0101
ln(SO_x)	–0.3218	0.132	–0.3641	–0.5216	–0.2157
ln(CO_2/L)	0.4161	–0.5348	–0.1485	–0.0358	0.1259
ln(CH_4/L)	0.167	–0.4415	–0.2129	–0.2132	0.0148
ln(NO_x/L)	–0.0798	–0.1418	–0.352	–0.4745	0.0429
ln(SO_x/L)	–0.2481	–0.046	–0.3766	–0.5034	–0.2039
ln(VA/L)	1	–0.5003	–0.3698	–0.1868	0.0764
ln(L)	–0.5003	1	0.0282	–0.1351	–0.0837
ln(TO$_{EU15}$)	–0.3698	0.0282	1	0.7964	–
ln(TO$_{extraEU15}$)	–0.1868	–0.1351	0.7964	1	–
ln(R&D/VA)	0.0764	–0.0837	–	–	1

Notes: Correlation between panel variables is equal to $corr(x_{it}, y_{it})=(\beta_1*\beta_2)^{1/2}$, with β_1 and β_2 given by REM estimates of equations $y_{it}=\alpha+\beta_1 xi_t+v2i_t$ and $x_{it}=\alpha+\beta_2 y_{it}+v2_{it}$.

other leading countries in terms of policy efforts. Future assessments, e.g. of the EU ETS scheme operative since 2005 in the EU (Alberola *et al.* 2008, in press; Smith and Swierzbinski 2007) would provide subjects for further research. The evidence is nevertheless as expected and, in part (in addition to the main sources of private transport and household emissions), a reason for the lack of absolute delinking regarding GHGs in the Italian economy so far.

Moving on to the results for trade openness and R&D (all estimated regressions omit services), we note that trade openness dynamics do not emerge as significant. This may be due to the compensating and opposite effects (pollution haven and capital abundance) of trade on emissions, an explanation proposed by several authors who also found not (very) significant relationships.

R&D[12] overall is not relevant: a 10 per cent statistical significance emerges for carbon, with a positive partially counterintuitive sign, which may reflect the weak eco-innovation content of and low environmental expenditure on process innovation dynamics in Italian industries, on average. Economic significance is also low, and the coefficient is negligible.

CO_2, STIRPAT specifications

In this type of analysis we refer to effects on emissions *per se*, not emissions technical efficiency, as stated. Table 5.5 sums up the main regressions related to comments in the text.

First, we can see that relative delinking is confirmed for carbon.

Looking at the evidence for industry and services, relative and absolute delinking respectively are generally confirmed by the STIRPAT models. The

Table 5.3 Descriptive statistics

VARIABLE	*MIN*	*MAX*	*MEAN (overall)*	*MEDIAN (overall)*
Whole NAMEA [29 branches] (1990–2005)				
VA/L	11.51 *(A 1990)*	528.5 *(CA 2000)*	62.83	41.51
L	6 *(CA 2000, CA 2001)*	3660 *(G 1991)*	780.64	479.5
Industry [C, D, E, F] (1990–2005)				
VA/L	22.44 *(DD 1990)*	528.5 *(CA 2000)*	74.81	45.22
L	6 *(CA 2000, CA 2001)*	1898 *(F 2005)*	379.07	266
Services [G, H, I, J, K, L, M, N, O] (1990–2005)				
VA/L	27.11 *(H 2004)*	111.53 *(K 1990)*	48.83	39.07
L	588 *(J 2000)*	3660 *(G 1991)*	1575.58	1450
Trade openness [A, B, C, D, E] (1995–2004)				
VA/L	15.63 *(A 1995)*	528.5 *(CA 2000)*	74.69	46.07
L	6 *(CA 2000, CA 2001)*	1640 *(A 1995)*	348.62	253
TO EU_{15}	0.0241 *(E 1999)*	4.8922 *(DF 2003)*	1.0334	0.7805
TO extra EU_{15}	0.0442 *(E 1997)*	12.796 *(DF 2002)*	1.3614	0.7368
Research & development [D, E, F] (1991–2003)				
VA/L	23.76 *(DD 1991)*	321.8 *(DF 1995)*	59.67	44.79
L	24 *(DF 2002, DF 2003)*	1794 *(F 2003)*	421.1	279
R&D/VA	0.0408 *(F 1992)*	179.7654 *(DM 1993)*	20.999	3.2635

main evidence from the STIRPAT framework relates to the emissions–labour relationship, which is implicitly defined in the EKC model. We note first that, on average at least, the employment trend, as in other countries, is decreasing for industry and increasing for services over the period considered. We focus on the specific figures for industry and services which we believe are more relevant than aggregate estimates. For industry, the elasticity is positive (0.5 for carbon).

On the basis of the empirical evidence, in the period considered we can propose a labour-saving interpretation: emissions decrease less than employment in industry, which has 'destroyed' labour. On the other hand, the employment

Table 5.4 EKC models for CO_2

	EKC 1 *[aggr]*	*EKC 2* *[ind]*	*EKC 3* *[serv]*	*EKC 5* *[R&D/VA]*
	ln(CO_2/L)	*ln(CO_2/L)*	*ln(CO_2/L)*	*ln(CO_2/L)*
ln(VA/L)	0.4727***	0.4965***	–7.4279***	0.4726***
ln(VA/L)^2			0.9458***	
ln(R&D/VA)				0.0261*
Kyoto	0.0416*** (104.25%)	0.1098*** (111.6%)	–0.0583** (94.34%)	0.0818*** (108.52%)
Constant	7.4618***	7.9769***	22.4173***	8.0405***
R^2 (overall)	0.2935	0.449	0.036	0.6309
F test	57.12***	97.32***	15.81***	29.08***
F test fixed effects	1904.22***	1863.46***	1079***	1018.55***
N*T	464	288	144	208
Period	1990–2005	1990–2005	1990–2005	1991–2003
Turning point	–	–	50.743*** *(74)*	–
Shape (VA/L)	Linear relationship	Linear relationship	U-shape	Linear relationship

Notes: Coefficients are shown in cells (*** p-values < 0.01, ** p-values < 0.05, * p-values < 0.1). All specifications are estimated using FE model; individual fixed effects coefficients are not shown for brevity. Kyoto coefficient: between brackets, average emissions in 1998–2005 given 1990–1997 average equal to 100 per cent are shown. TP: between brackets there is the percentile of each TP.

increases in services tend to be associated with 'emissions-saving' dynamics. This evidence should hold also for the future when we would expect similar trends, although probably mitigated in terms of its relative size.

Finally, trade openness (coverage 1995–2004) is negatively related (though the size is negligible for CO_2, and for both EU_{15} and extra-EU_{15} trade dynamics). This can be interpreted as the pollution-haven effect, which is generally driven by trade openness, being more significant if we focus on emissions rather than emissions efficiency. This suggests an area for future research. Given that trade openness in the extra-EU_{15} has increased since 1999, the higher elasticity we estimate has some serious implications for future scenarios. R&D is again positively but negligibly driving carbon emissions.

SO_x, EKC specifications

The EKC related evidence suggests non-linear cubic relationships which are worthy of careful investigation. Table 5.6 presents the main regressions in relation to comments in the text.

The estimated SO_x inverted N leaves at the right of the second TP a bunch of few units. A U-shape then becomes plausible. The coherence of this evidence with the very sharp decrease in emissions over the last 20 years can be found in the

Table 5.5 STIRPAT models for CO_2

	STIRPAT 1 [aggr]	*STIRPAT 2 [ind]*	*STIRPAT 3 [serv]*	*STIRPAT 4a [TO_{EU15}]*	*STIRPAT 4b [$TO_{extraEU15}$]*	*STIRPAT 5 [R&D/VA]*
	$ln(CO_2)$	*$ln(CO_2)$*	*$ln(CO_2)$*	*$ln(CO_2)$*	*$ln(CO_2)$*	*$ln(CO_2)$*
ln(VA/L)	0.1555**	0.3236***	–44.28546**	–3.7014**	–3.665**	0.3106***
$ln(VA/L)^2$			10.8242**	0.9108**	0.9052**	
$ln(VA/L)^3$			–0.8807**	–0.0674**	–0.0671**	
ln(L)	0.2742***	0.5042***	19.9504***	0.4131***	0.3848***	0.3914***
$ln(L)^2$			–1.3123***			
$ln(TO_{EU15})$				–0.0801**		
$ln(TO_{extraEU15})$					–0.0857**	
ln(R&D/VA)						0.0219*
Kyoto	0.0536*** (105.51%)	0.0985*** (110.35%)	–0.055* (94.65%)	0.0491*** (105.03%)	0.055*** (105.65%)	0.0759*** (107.89%)
Constant	13.0296***	11.3477***	0.1075	17.6752***	17.7354***	12.1557***
R^2 (overall)	0.0938	0.1471	0.0812	0.3513	0.3669	0.0151
F test	13.77***	30.31***	13.40***	7.76***	7.79***	12.98***
F test fixed effects	1797.36***	2056.23***	1243.94***	1162.54***	1093.39***	1071.92***
N*T	464	288	144	190	190	208
Period	1990–2005	1990–2005	1990–2005	1995–2004	1995–2004	1991–2003
Turning point(s) (VA/L)	–	–	51.655*** *(74)* 70.052*** *(79)*	22.1*** *(9)* 369.753*** *(97)*	21.656*** *(8)* 373.719*** *(97)*	–
Turning point (L)	–	–	2000.333*** *(84)*	–	–	–
Shape (VA/L)	Linear relationship	Linear relationship	Inverted N-shape	Inverted N-shape	Inverted N-shape	Linear relationship

Notes: Coefficients are shown in cells (significance at 1%, 5% and 10% is denoted by ***, ** and * respectively). All specifications are estimated using FE model; individual fixed effects coefficients are not shown for brevity. Kyoto coefficient: between brackets, average emissions in 1998–2005 given 1990–1997 average equal to 100 per cent are shown. TP(s) (two for cubic and one for quadratic specifications): between brackets there is the percentile of each TP.

Table 5.6 EKC models for SO_x

	EKC 1 [aggr]	*EKC 2 [ind]*	*EKC 3 [serv]*	*EKC 4a [TO_{EU15}]*	*EKC 4b [$TO_{extraEU15}$]*	*EKC 5 [R&D/VA]*
	$ln(SO_x/L)$	*$ln(SO_x/L)$*	*$ln(SO_x/L)$*	*$ln(SO_x/L)$*	*$ln(SO_x/L)$*	*$ln(SO_x/L)$*
ln(VA/L)	–11.1397***		–19.8059***			–0.4014*
$ln(VA/L)^2$	2.4244***		2.5326***			
$ln(VA/L)^3$	–0.1691***					
$ln(TO_{EU15})$				–0.5588***		
$ln(TO_{extraEU15})$					–0.6415***	
ln(R&D/VA)						–0.1427***
Kyoto	–1.224*** (29.41%)	–1.2114*** (29.78%)	–1.0875*** (33.71%)	–0.9168*** (39.98%)	–0.864*** (42.15%)	–0.9923*** (37.07%)
Constant	19.2953***	3.7807***	39.3315***	3.2841***	3.1642***	5.6374***
R^2 (overall)	0.0745	0.0589	0.0291	0.0355	0.0204	0.0174
F test	218.81***	704.57***	64.84***	209.14***	213.68***	223.3***
F test fixed effects	342.92***	646.16***	127.48***	633.84***	647.78***	248.36***
N*T	464	288	144	190	190	208
Period	1990–2005	1990–2005	1990–2005	1995–2004	1995–2004	1991–2003
Turning point(s)	46.581*** *(65)* 303.881*** *(98)*	–	49.908*** *(74)*	–	–	–
Shape (VA/L)	Inverted N-shape	No significant relationship	U-shape	No significant relationship	No significant relationship	Linear relationship

Notes: Coefficients are shown in cells (*** p-values < 0.01, ** p-values < 0.05, * p-values < 0.1). All specifications are estimated using FE model; individual fixed effects coefficients are not shown for brevity. Kyoto coefficient: between brackets, average emissions in 1998–2005 given 1990–1997 average equal to 100 per cent are shown. TP (two for cubic and one for quadratic specifications): between brackets there is the percentile of each TP.

statistically and economically very significant Kyoto break factor. For example, this seems to capture all the statistical explanatory power in the SO_x industry regression. In other words, we can say that for SO_x the role played by income dynamics is less relevant for explaining environmental pressures relative to more exogenous factors that are only partly captured here. These include the many regulatory interventions on air pollution by the EU since the early 1980s (e.g. Directive 1980/779/EC substituted by 1999/30/EC, the Directive 1999/32/EC, the new CAFE (Clean Air for Europe) programme from 2005), and the adoption of end-of-pipe technologies which are currently the main tool for addressing pollution.

For services, we show U curves mainly depending on the J and K outlier dynamics, already commented on above. In addition, services shows the expected negative linear income–environment dynamics, well beyond the EKC turning point. It remains relevant to assess the extent to which stagnation periods may affect, more or less substantially, the structural trend depicted by the EKC hypothesis.

Trade openness shows negative and significant coefficients. If on the one hand the extra-EU_{15} related evidence suggests a stronger weight of the 'pollution-haven' factor relative to endowments, on the side of EU_{15} trade the motivations may include a number of perspectives. First, increasing trade openness (38 per cent and 65 per cent from 1995 to 2004 respectively for the EU_{15} and extra-EU_{15}) is associated with a stricter integration in terms of environmental policy, which may explain the good and converging performance of eastern newcomers since the late 1990s (Zugravu *et al.* 2008). We can confirm that Italy is a follower and a convergent country in terms of environmental policy implementation in the EU context, thus this hypothesis has robust roots. Such convergence may also (have) occur(red) along pure market dynamics through technological spillovers and increasing technological and organisational environmental standards in order to compete with European leaders. Second, along the path of increasing openness, intra-branch specialisations over time may be favouring more efficient technologies and production processes. This would support increasing Italian specialisation in more environmentally benign sectors and production processes. It is obvious that a structural decomposition analysis would be the best tool for assessing the relevance of these driving forces captured here, at a lower level of sector detail, using econometric techniques that result in more 'average trends and statistical regularities'.

Finally, it should be noted that for SO_x R&D emerges as being associated with a statistically negative and economically significant coefficient.[13]

SO_x, STIRPAT specifications

As far as the evidence of emissions-labour productivity is concerned, the results confirm the EKC analyses (Table 5.7). For SO_x in the aggregate, the same comments on CA and DF apply as in Marin and Mazzanti (2009); for services we note again the need for an investigation of sector specificity: sector I explains the

Table 5.7 STIRPAT models for SO_x

	STIRPAT 1 [aggr]	*STIRPAT 2 [ind]*	*STIRPAT 3 [serv]*	*STIRPAT 4a [TO_{EU15}]*	*STIRPAT 4b [$TO_{extraEU15}$]*	*STIRPAT 5 [R&D/VA]*
	ln(SO_x)	*ln(SO_x)*	*ln(SO_x)*	*ln(SO_x)*	*ln(SO_x)*	*ln(SO_x)*
ln(VA/L)	–2.2639***		–2.1298***	3.8154***	4.0174***	–0.6676***
$\ln(VA/L)^2$	0.2273**			–0.4424***	–0.4591***	
$\ln(VA/L)^3$						
ln(L)	0.7889***	0.9916***	–2.8852***	–3.6036***	–3.8822***	
$\ln(L)^2$				0.5108***	0.5093***	
$\ln(TO_{EU15})$				–0.7244***		
$\ln(TO_{extraEU15})$					–0.8058***	
ln(R&D/VA)						–0.1497***
Kyoto	–1.2064*** (29.93%)	–1.2123*** (29.75%)	–0.7356*** (47.92%)	–0.914*** (40.09%)	–0.8565*** (42.46%)	–1.0019*** (36.72%)
Constant	9.3382***	3.8264**	37.121***	4.5337	5.3881	12.3995***
R^2 (overall)	0.0457	0.1329	0.05	0.0445	0.0747	0.0167
F test	225.23***	386.51***	64.67***	83.29***	85.13***	242.5***
F test fixed effects	319.81***	609.69***	165.01***	415.78***	422.36***	244.51***
N*T	464	288	144	190	190	208
Period	1990–2005	1990–2005	1990–2005	1995–2004	1995–2004	1991–2003
Turning point (VA/L)	145.5714* *(92)*	–	–	74.570*** *(79)*	79.493*** *(82)*	–
Turning point (L)	–	–	–	34.042** *(14)*	45.214*** *(16)*	–
Shape (VA/L)	U-shape	No significant relationship	Linear relationship	Inverted U-shape	Inverted U-shape	Linear relationship

Notes: Coefficients are shown in cells: (*** p-values < 0.01, ** p-values < 0.05, * p-values < 0.1). All specifications are estimated using FE model; individual fixed effects coefficients are not shown for brevity. Kyoto coefficient: between brackets, average emissions in 1998–2005 given 1990–1997 average equal to 100 per cent are shown. TP: between brackets there is the percentile of each TP.

N-shape, which is transformed into a linear negative dynamics when the sector is omitted and shows a weaker delinking with respect to other sectors. The evidence for Kyoto factors, trade openness and R&D are the same as for the EKC analysis.

Conclusions

The results show that, looking at sector evidence, both decoupling and also eventually re-coupling trends could emerge along the path of economic development. Both the way that the stagnation periods affect environmental performance and contingent sector specificity emerge as relevant explanations of the various U- and N-shapes. CO_2 seems still to be associated only with relative delinking. Methane is moderately decreasing in recent years (analyses not shown here), but, being a minor gas, the overall performance for GHGs is not compliant with the Kyoto targets, which do not appear to have generated a structural break in the dynamics, at least for GHGs. SO_x presents decreasing patterns, and the shape is affected by time non-observed effects rather than income, such as possible exogenous innovation and policy-related factors. Services tend to show stronger delinking patterns across emissions. Trade expansion validates the pollution haven in some cases, but also shows negative signs when EU_{15} trade only is considered: this may be due to technology spillovers and a positive *race to the top* rather than to the bottom among the EU_{15} trade partners (Italy and Germany as main exporters and also trade partners in the EU). Finally, general R&D expenditures show weak correlation to emissions efficiency.

EKC and IPAT derived models provide similar conclusions overall; the emissions–labour elasticity estimated in the latter is generally different from one, suggesting in most cases, and for both services and industry, a scenario characterised by emissions-saving technological dynamics (as well as labour saving in relation to GHGs in industry).

The application of heterogeneous panel estimators is a direction for future applied research to assess the extent to which U- and N-shapes emerge from average trends. Both sector-based and country-based analyses would benefit from a comparison of homogenous and heterogeneous panel estimations. From a data construction point of view, future research should aim at using environmental R&D and innovation data at sector level; a final and challenging research direction would be to set up trade factors in terms of inter-sector and intra-sector datasets, by exploiting I-O tables and NAMEA or other compatible sources related to trading partners.

Notes

1 STIRPAT is Stochastic Impacts by Regressions on Population, Affluence and Technology.
2 EEA (2004), Markandya and Rübbelke (2003), Pearce (1992, 2000) and Barker and Rosendahl (2000).
3 $\delta_3 \ln(E_{it}/VA_{it})$ enters the residuals.

4 Mazzanti and Zoboli (2009), Stern (2004), Berndt and Wood (1979); Koetse *et al.* (2008).
5 Ike (1999), Vaze (1999), de Haan and Keuning (1996) and Keuning *et al.* (1999), among others, provide descriptive and methodological insights on NAMEA for some of the major countries. Steenge (1999) provides an analysis of NAMEA with reference to environmental policy issues.
6 For an exhaustive overview of environmental accounting systems see the so-called SEEA 2003 (UN *et al.* 2003).
7 Import, export and trade openness respectively, with partners inside and outside the EU_{15} area.
8 ANBERD is Analytical Business Enterprise Expenditure on Research and Development.
9 International Standard Industrial Classification (http://unstats.un.org/).
10 The main fact is that K shows decreasing labour productivity, due to the high growth of employment in services and in some sectors such as K. Employment growth is then higher than value added growth; given that emission efficiency increases, the result is a positive sign captured by panel estimates. Heterogeneous panel models may circumvent this estimation output. This example shows the importance of investigating latent sector dynamics, and the relevance of analysing the driving forces of decoupling and recoupling trends.
11 The Italian government obtained exemption from payment of emissions quotas for industrial sectors producing paper (DA), pottery, glass (DI) and steel (DJ). The test of the EKC model separately for those branches highlights the bad performance of paper (elasticity above unity), a smaller delinking in comparison with the industry for pottery and glass (elasticity lower than 1 but higher than 0.5) and a robust absolute delinking for steel.
12 The correlation between R&D/VA and VA/L is low: 0.0764. We find a significant and positive relationship between economic productivity (VA/L) and innovative efforts (R&D/VA) only after the third lag (in years) of the R&D/VA ratio.
13 If this perhaps seems to conflict with what was said above about end-of-pipe exogenous effects associated with technological expenditures external to the firm (acquisition of equipment, thus not core R&D), in house or external R&D may be better and primarily targeted towards externalities such as SO_x, embodying some element of privateness in terms of rent appropriability, compared to pure public goods such as GHGs. It remains true that the movement towards integrated process innovations away from end-of-pipe solutions entails an higher degree of complementarity between GHGs and pollutant reductions. Similar evidence is found by Vollebergh *et al.* (2009): SO_x is determined more by technology and other somewhat exogenous events, including policies not completely captured by the income dynamics.

References

Alberola, E., Chevallier, J. and Chèze, B. (2008) 'Price drivers and structural breaks in European carbon prices 2005–2007', *Energy Policy*, 36: 787–97.

Alberola, E., Chevallier, J. and Chèze, B. (in press) 'The EU emissions trading scheme: Country-specific perspectives of the impacts of industrial production on carbon prices', *Journal of Policy Modelling*, 31: 446–62.

Barker, T. and Rosendahl, K.E. (2000) 'Ancillary Benefits of GHG Mitigation in Europe: SO_2, NO_x and PM_{10} reductions from policies to meet Kyoto targets using the E3ME model and EXTERNE valuations, Ancillary Benefits and Costs of Greenhouse Gas Mitigation', Proceedings of an IPCC Co-Sponsored Workshop, March, Paris: OECD.

Berndt, E.R. and Wood, D.O. (1979) 'Engineering and econometric interpretations of energy-capital complementarity', *American Economic Review*, 69: 342–54.

Cole, M.A. (2003) 'Development, trade and the environment: How robust is the environmental Kuznets curve?' *Environment and Development Economics*, 8: 557–80.

Cole, M.A. (2004) 'Trade, the pollution haven hypothesis and the environmental Kuznets curve: examining the linkages', *Ecological Economics*, 48: 71–81.

Cole, M.A. (2005) 'Re-examining the pollution–income relationship: a random coefficients approach', *Economics Bulletin*, 14: 1–7.

Cole, M.A. and Elliott, R. (2003) 'Determining the trade–environment composition effect: the role of capital, labor and environmental regulation', *Journal of Environmental Economics and Management*, 46: 363–83.

Cole, M.A. and Neumayer, E. (2004) 'Examining the impact of demographic factors on air pollution', *Population & Environment*, 26: 5–21.

Copeland, B.R. and Taylor, M.S. (2004) 'Trade, growth and the environment', *Journal of Economic Literature*, 42: 7–71.

de Haan, M. and Keuning, S.J. (1996) 'Taking the environment into account: the NAMEA approach', *Review of Income and Wealth*, 42: 131–48.

Dietz, T. and Rosa, E.A. (1994) 'Rethinking the environmental impacts of population, affluence and technology', *Human Ecology Review*, 1: 277–300.

Dietzenbacher, E. and Mukhopadhay, K. (2006) 'An empirical examination of the pollution haven hypothesis for India: towards a green Leontief paradox?', *Environmental & Resource Economics*, 36: 427–49.

EEA (2004) 'Air Pollution and Climate Change Policies in Europe: Exploring Linkages and the Added Value of an Integrated Approach', EEA Technical report No. 5/2004, Copenhagen: European Environment Agency.

Femia, A. and Panfili, P. (2005) 'Analytical applications of the NAMEA', Paper presented at the annual meeting of the Italian Statistics Society, Rome: Italian Statistics Society.

Frankel, J. and Rose, A. (2005) 'Is trade good or bad for the environment?', *Review of Economics and Statistics*, 87: 85–91.

Ike, T. (1999) 'A Japanese NAMEA', *Structural Change and Economic Dynamics*, 10: 122–49.

Keuning, S., van Dalen, J. and de Haan, M. (1999) 'The Netherlands' NAMEA; presentation, usage and future extensions', *Structural Change and Economic Dynamics*, 10: 15–37.

Koetse, M.J., de Groot, H.L.F. and Florax, R.J.G.M. (2008) 'Capital-energy substitution and shifts in factor demand: A meta analysis', *Energy Economics* 30: 2236–51.

Marin, G. and Mazzanti, M. (2009) 'The dynamics of industrial emissions: the role of productivity, trade and R&D', *Journal of Innovation Economics*, 2: 91–117.

Markandya, A. and Rübbelke, D.T.G. (2003) 'Ancillary Benefits of Climate Policy', FEEM working paper 105, Milan: FEEM.

Martinez-Zarzoso, I., Bengochea-Morancho, A. and Morales-Lage, R. (2007) 'The impact of population on CO_2 emissions: evidence from European countries', *Environmental and Resource Economics*, 38: 497–512.

Mazzanti, M. and Marin, G. (2009), 'Emissions Trends and Labour Productivity Dynamics Analyses of De-coupling/Recoupling on a 1990–2005 NAMEA', FEEM working paper 50, Milan: FEEM.

Mazzanti, M. and Zoboli, R. (2005), 'Waste indicators, economic drivers and environmental efficiency: perspectives on delinking and empirical evidence for Europe', *Environmental Sciences*, 2: 409–25.

Mazzanti, M. and Zoboli, R. (2009) 'Environmental efficiency and labour productivity: trade-off or joint dynamics?', *Ecological Economics*, 38: 1182–94.

Mazzanti, M., Montini, A. and Zoboli, R. (2008a) 'Environmental Kuznets curve for air pollutant emissions in Italy: evidence from environmental accounts NAMEA panel data', *Economic System Research*, 20: 279–305.

Mazzanti, M., Montini, A. and Zoboli, R. (2008b) 'Environmental Kuznets curves and greenhouse gas emissions. Evidence from NAMEA and provincial data', *International Journal of Global Environmental Issues*, 8: 392–424.

Pearce, D.W. (1992) 'Secondary Benefits of Greenhouse Gas Control', CSERGE working paper 12, London: UCL.

Pearce, D.W. (2000) 'Policy Framework for the Ancillary Benefits of Climate Change Policies', Paper presented at the IPCC Workshop on Assessing the Ancillary Benefits and Costs of Greenhouse Gas Mitigation Strategies, Washington DC: IPCC.

Smith, S. and Swierzbinski, J. (2007) 'Assessing the performance of the UK Emission Trading Scheme', *Environmental and Resource Economics*, 37: 131–58.

Steenge, A. (1999) 'Input–output theory and institutional aspects of environmental policy', *Structural Change and Economic Dynamics*, 10: 161–76.

Stern, D. (1998) 'Progress on the environmental Kuznets curve?', *Environment and Development Economics*, 3: 173–96.

Stern, D. (2004) 'Elasticities of Substitution and Complementarity', *Rensselaer Working Papers in Economics*, No. 20/2004.

UN, Eurostat, IMF, OECD and World Bank (2003) *Integrated Environmental and Economic Accounting 2003 – Handbook of National Accounting*, Final Draft circulated for information prior to official editing, http://unstats.un.org/.

Vaze, P. (1999) 'A NAMEA for the UK', *Structural Change and Economic Dynamics*, 10: 99–121.

Vollebergh, H., Melenberg, B. and Dijkgraaf, E. (2009) 'Identifying reduced-form relations with panel data: The case of pollution and income', *Journal of Environmental Economics and Management*, 58: 27–42.

Wooldridge, J. (2005) *Introductory Econometrics – A Modern Approach*, Mason, OH: Thomson South-Western, 2nd edition.

York, R., Rosa, E.A. and Dietz, T. (2003) 'STIRPAT, IPAT and ImPACT: analytic tools for unpacking the driving forces of environmental impacts', *Ecological Economics*, 46: 351–65.

Zugravu, N., Millock, K. and Duchene, G. (2008) 'The Factors Behind CO_2 Emission Reduction in Transition Economies', FEEM working paper 50, Milan: FEEM.

6 Production-related air emissions

A decomposition analysis for Italy

Aldo Femia and Renato Marra Campanale[1]

Introduction

This chapter decomposes the Air Emission Accounts from the Italian National Accounting Matrix including Environmental Accounts (NAMEA) in order to investigate the influence of economic activities on three environmental issues: greenhouse effect, acidification, and the formation of ozone in the troposphere.[2]

The data, which are based on official environmental accounting statistics, are suited to the application of decomposition techniques at the meso-aggregation (branch of economic activity) level. Air Emission Accounts, a system of hybrid satellite national accounts, use the same concepts, definitions and classifications as national accounts, which guarantees that different types of data, in this case environmental pressure and economic, are correctly combined in the NAMEA, as they refer to the same activities.[3] Air Emission Accounts data are also consistent with the Italian national Air Emission Inventory, on which they are based, and which are reconciled with national accounting conventions (application of economic boundaries including the residence principle) and provide a different breakdown of total emissions based on classification of economic activities rather than technical processes.

Our analysis also draws on data from Energy Use Accounts. These include data on industry combustion of fuels causing air emissions, but exclude fuel transformation and utilization as raw material. The fuels considered are coal, coke gas and other gases, coal coke, fuel oil, natural gas, motor gasoline, waste, diesel oil, jet fuel and kerosene, lignite, peat, other petroleum products. The collection and processing of these data is closely connected to the compilation of the National Accounts and Air Emission Accounts, which guarantees consistency between energy data and economic and environmental data.

In terms of economic variables, it should be noted that a meaningful connection between the changes in these and environmental variables over time is possible only if the changes in economic data are measured such that they are not influenced by price changes. Our analysis, therefore, uses chained measures of economic changes in terms of volume, which compare the output for one year, evaluated at that year's (current) prices with the output of the next year evaluated at the previous year's prices.

We decompose aggregate air emissions for: (i) greenhouse gases (GHG): the emissions that give rise to a greenhouse effect are carbon dioxide (CO_2), nitrous oxide (N_2O) and methane (CH_4);[4] the index for the GHG theme is given by the weighted sum in CO_2-equivalents of these emissions; (ii) acidification: nitrogen oxide (NO_x), sulphur dioxide (SO_2) and ammonia (NH_3) emissions contribute to soil acidification; the index for this theme is given by sum of their Potential Acid Equivalents (PAE);[5] (iii) Tropospheric Ozone Precursors: nitrogen oxides (NO_x), non-methane volatile organic compounds (NMVOC), carbon monoxide (CO) and methane (CH_4); their weighted emissions produce the Tropospheric Ozone Forming Potentials (TOFP) index.[6]

Table 6.1 shows changes in the theme indices over the period 1992–2006.

Figure 6.1 depicts the emissions indices for the three environmental themes for the period 1992–2006 in relation to the two economic indicators of output and value added. The time series shows strong growth of the economic variables; air emissions show two different trends. GHG have grown, although at a slower rate

Table 6.1 Production-related greenhouse gases, acidification and tropospheric ozone precursor emissions in Italy in 1992, 2006 and per cent change 1992–2006 (tons CO_2-equivalents, tons PAE, and tons TOFP)

	1992	*2006*	*Change 1992–2006 (%)*
Greenhouse gases	430,781,051	454,771,116	5.6
Acidification	110,145	54,389	–50.6
Tropospheric ozone precursors	3,762,529	1,828,287	–51.4

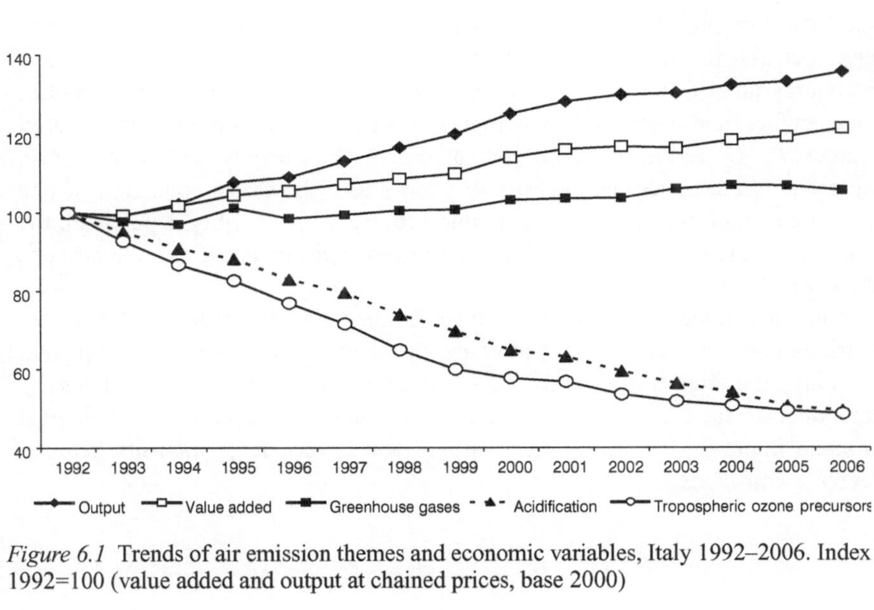

Figure 6.1 Trends of air emission themes and economic variables, Italy 1992–2006. Index 1992=100 (value added and output at chained prices, base 2000)

than the economy, showing therefore relative decoupling, while acidification and the tropospheric ozone precursors decreased, showing absolute decoupling.[7]

Clearly, economic growth has been accompanied by some structural changes, otherwise emissions would have increased proportionally to the volume of economic activity. What kinds of structural changes? Is the decoupling due to changes in the composition of economic activities, that is, a shift towards less polluting activities? If so, how much of the emissions containment is due to this phenomenon? Has the increased efficiency in the use of energy from fuels, that is, the ability to produce more output from a given fuel input, played a significant role? What has been the importance of changes in the composition of fuel inputs, that is, the shift towards less polluting fuels (e.g. natural gas replacing coal or oil)? We try to answer these questions by means of a decomposition analysis, and for each environmental theme, express the variation in the corresponding emissions index as the sum of the effects of the evolution of these underlying factors.

Our main equation decomposes into four effects, the industry-related air emissions change in Italy, between 1992 and 2006.[8] Emissions (P^t) in a given year can be written as:

$$P^t = \sum Y^t * \left(Y_i^t / Y^t\right) * \left(E_i^t / Y_i^t\right) * \left(P_i^t / E_i^t\right) \tag{6.1}$$

where Y is total national output, representing the level of economic activity: general economic growth is usually the main driver of increased emissions;[9] Y_i / Y is the share of industry i in total output, representing the structure of production: for any given total output level, different compositions of this output by economic activity – that is, the production of different goods and services – give rise to different total emissions; E_i / Y_i: is the energy from fuels used per output unit in industry i, measured in terajoule (TJ) per € million, representing the energy intensity of output: the vector of these intensities shows to what extent industries have been able to vary their production technologies towards more efficient use of the energy input from fuels, for example, by exploiting substitution possibilities between fuels and other productive inputs or between fuels and renewable energy sources; P_i / E_i are the emissions generated per unit of energy used in industry i, representing the emissions intensity of energy use: shows the emissions of one terajoule of fuel use in the i-th economic activity, which mainly depends on the mix of fuels used to obtain each TJ of energy and on the extent of use of emissions abating technologies.

Thus, the model used for the analysis in this paper (Equation 6.1) provides a strictly economic determinant of emissions change, a composition component reflecting the impact on emissions of structural changes, and two intensity components that correspond to technological choice. In general, technological progress improves the utilization of energy, and lower energy intensity helps to decrease emissions.

Results

Decomposition of GHG

In Italy the increase in GHG emissions due to economic activities was some 24 million tonnes in the period 1992–2006 (Table 6.1). Decomposition analysis of this change (Figure 6.2) shows that this increase is due exclusively to the 'economic growth' component. Indeed, had the effects of economic growth not been offset by other components the total change would have been of the order of 30.8 per cent rather than 5.6 per cent. This is a straightforward, almost obvious result: the volume effect, given the characteristics of the decomposition technique, cannot be very different from the volume growth. In our case the volume effect is somewhat lower than output growth (35.5 per cent), because the weight factors are not fixed at their initial value, but correspond to an average of all the values achieved throughout the whole period, which values have been generally constantly decreasing. It should be noted that all the other components acted to keep down the increase in emissions.

It is due mainly to the environmentally positive effect of the two technological components that the total change was only 5.6 per cent: an improvement in the emissions intensity of energy use, and the energy intensity of output would have led to a reduction of 20.7 per cent in potential emissions. The structure of the production component also played a significant, though less important role in terms of decreasing the GHG emissions (–4.5 per cent).

Figure 6.3 is similar to Figure 6.2, but depicts the importance of each component in year on year changes. The economic growth component provides for a positive change in GHG emissions in every year of the period except 1993. The structure

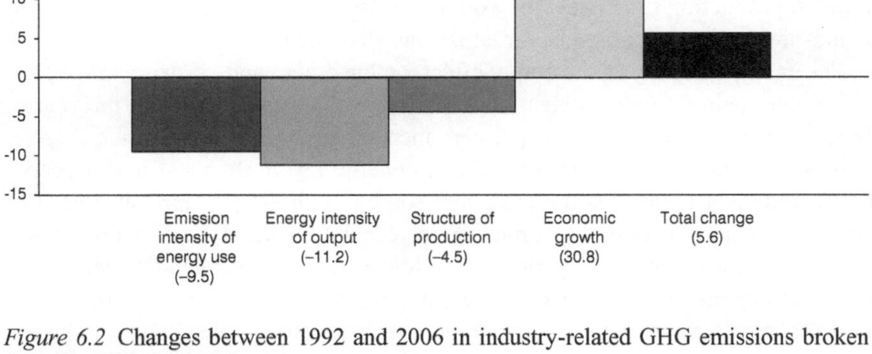

Figure 6.2 Changes between 1992 and 2006 in industry-related GHG emissions broken down by effect (percentage)

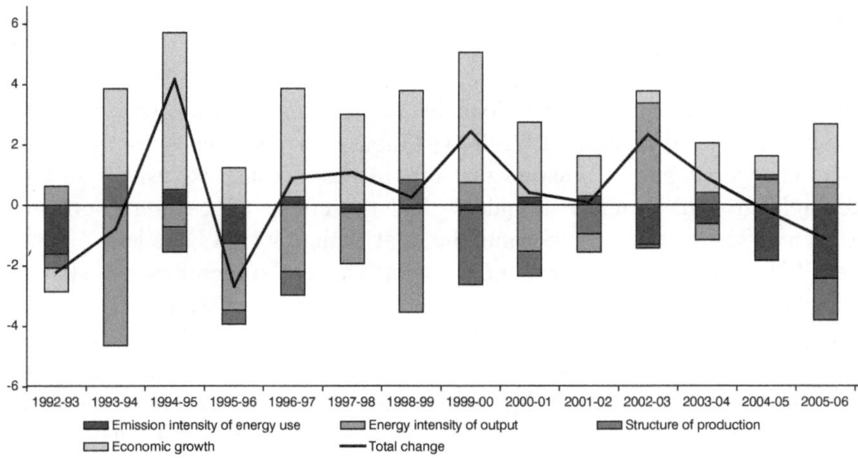

Figure 6.3 Year by year changes in industry-related GHG emissions by effect 1992–2006 (percentage of 1992 emissions)

of production component shows a generally favourable impact on total emissions variation; however, it is difficult to identify a pattern to this effect, which shows yearly swings.

The two intensity effects show different trends over the period. There is a quite steady annual change in GHG emissions intensity of energy use that is not evident in energy use intensity of output. Indeed, the 1992–2006 overall change (–9.5 per cent) attributable to the emissions intensity of energy use was achieved mostly by the cumulation of small year-by-year efficiency increases over the period (although the sign of the component is positive in only three years). This points out a real, albeit slow, amelioration of the environmental efficiency of Italian industry. Indeed, the fuel quantities in our energy use data are those actually used for energy purposes, so that lower volumes of gases are emitted per TJ of fuel energy, this is dependent on four kinds of changes: decrease in non-fuel dependent emissions (e.g. from solvent use); shift towards less polluting fuels; use of technologies that improve production processes, for example, integrated technologies; installation of end-of-the-pipe technologies for emissions abatement.

The energy intensity of the output effect, on the other hand, is characterized by an irregular pattern, with sudden and often large changes observed in the yearly data. In particular, we find a 3.4 per cent increase in 2003 and a positive change in 2005 and 2006, pointing to less efficient energy use in the most recent years. This component comprises the effects of some possible real improvements, for example, the introduction of less fuel intensive techniques and the discontinuation of energy-wasting machinery or plant. However, both its overall importance and volatility may be explained, at least in part, by the delocalization of Italian industrial production. In fact, more and more intermediate Italian industry products are manufactured abroad, with only the final stages being performed in Italy.[10] This implies that the output is obtained without the energy input by the 'producing' industry (which in fact buys in from abroad intermediate inputs, almost finished goods, and re-sells them after slight transformation).

The decomposition analysis highlights to what extent individual industries contribute to individual components. Here we can distinguish the most relevant industries, ranking them according to their importance as GHG polluters, to show how they perform with regard to the four decomposition components.

Table 6.2 shows the results for 11 industries, which, considered as a whole, as well as being responsible for most of the emissions, account for most of the change in total emissions (87.5 per cent of the change). The picture is quite varied, with some industries providing a total contribution towards the reduction of GHG emissions in absolute terms. The figures in Table 6.2 are normalized by being expressed as a percentage of 1992 total production emissions, so that were all industries included, the sum of the total changes by industry would be the 5.6 per cent reported in Figure 6.2.

GHG emissions from the electricity, gas and water industries account for almost 30 per cent of total Italian GHG emissions from production; the Air Emission Accounts assign emissions to the industries that produce them directly (responsibility of the producer approach). It should be noted that the total GHG emissions increase in these industries are higher than the Italian average because the emissions intensity of energy use did not change over the period (in the specialized energy production industry there was no significant technological change in the direction of emissions abatement per unit of energy input from fuels; there was only a reduction in the input of energy from fuels per unit of output). The structure of the production effect has the same sign as the volume effect, because the industry is characterised by an emissions intensity per output unit that is higher than the average, and grew more than the average.[11]

Also in 'Agriculture, hunting and forestry' the emissions intensity of energy use component does not change, but total change for the industry shows improvement, with emissions decreasing over the period by up to one per cent of total Italian production emissions in 1992. This change was largely due to the structure of production component: output in the agriculture sector, which is characterized by high energy inputs and emissions intensity, decreased sharply in 1992 to 2006 losing more than 40 per cent of its share in total output.

The manufacturing industries included in Table 6.2 show different performances. We note a significant worsening of GHG emissions only in 'Manufacture of other non-metallic mineral products' (providing a contribution of about two per cent) which is also due to a higher emissions intensity of energy use in 2006 (this component accounts for 1.5 per cent). The 'Manufacture of chemicals and chemical products' and 'Manufacture of basic metals' industries record a decrease in the total emissions due to technological factors; a remarkable improvement in the emissions intensity of energy use allows the reduction in NACE 24, while in 'Manufacture of basic metals' both intensity components combine to bring about the -1.1 per cent contribution to total change. In 'Manufacture of coke, refined petroleum products and nuclear fuel', the contribution of the scale component

Table 6.2 Decomposition of the change in GHG emissions of the most polluting industries between 1992 and 2006 (as percentage of the total amount of GHG emissions emitted by all industries in 1992)

Industries (NACE Rev. 1.1)		*Economic growth*	*Structure of production*	*Energy intensity of output*	*Emission intensity of energy use*	*Total change*
40–41	Electricity, gas and water supply	9.0	1.5	–3.8	–0.1	6.7
01–02	Agriculture, hunting and forestry	3.5	–3.1	–1.3	0.03	–0.9
26	Manufacture of other non-metallic mineral products	3.0	–0.7	–1.9	1.5	1.9
60–63	*Transport*	2.5	1.2	–1.0	–1.3	1.4
24	Manufacture of chemicals and chemical products	1.8	0.05	–0.1	–4.0	–2.3
27	Manufacture of basic metals	1.6	–0.2	–1.6	–0.9	–1.1
23	Manufacture of coke, refined petroleum products and nuclear fuel	1.7	–1.7	–0.6	0.5	–0.1
90–91–93	Sewage and refuse disposal, sanitation and similar activities; Activities of membership organisations; other services	1.5	–0.9	3.7	–3.6	0.7
51–52	*Trade*	1.2	–0.2	–2.4	–0.1	–1.4
17	Manufacture of textiles and textile products	0.7	–0.6	–0.3	–0.01	–0.3
15–16	Manufacture of tobacco, food products and beverages	0.6	–0.1	0.5	–0.7	0.3

to emissions growth is offset by the lower share in total output of this economic activity, meaning that in fact this industry has not grown; also its other components do not show significant change. NACE 15–16 and NACE 17 do not make a significant contribution to total emissions change in any of their components.

GHG emissions in transport activities increased between 1992 and 2006, providing a contribution of 1.4 per cent to overall emissions from production in 1992, since the negative contribution of the two intensity factors was not large enough to offset the two economic components. In NACE 90, 91 and 93, where GHG emissions stem mainly from waste treatment and disposal activities (NACE 90), the two intensity components cancel each other out, which results in a 0.7 per cent total emissions change. Finally, the -2.4 per cent change in "energy intensity of output" allowed a reduction in GHG emissions in the trade industries.

The industry disaggregation level has an important impact on the results. Using a different level of aggregation implies a different distribution of the total change between the intensity-related and the structural effects; the latter loses its importance when the industries are aggregated. Indeed, aggregation implies loss of information on the composition of the economy by industry, as the shift from one industry to another is not visible when they are both part of the same aggregate. At one extreme, we have the whole economy as a single industry, and therefore no structural effect: what in reality is a shift between activities appears as an efficiency gain (or loss). Aggregation does not affect the economic growth driver (which is the same for all industries however aggregated) and the total emissions change also does not depend on the aggregation. Therefore, the more aggregated the analysis, the more that intensity-related factors absorb the composition effect, even to the point that the latter may have the opposite sign. The composition effect can only be revealed by slicing up the economy: ideally pushed at the other extreme, where every change in production technique is a change in product, disaggregation would lead to all technological changes being described as structural changes, so that the intensity factors would lose their importance. To summarize then, the aggregation issue should be kept in mind when the results of decomposition analysis are being used for policy recommendations: efficiency gains could be overestimated since an examination of the industries in greater detail would reveal that some improvements are in reality the effects of the shift from heavily polluting productions to less polluting ones, within the same industry.

By comparing the above results with those obtained with an 8-industry disaggregation on the same data, we can provide an example of this phenomenon. Table 6.3 shows that in the 8-industry analysis the importance of the more efficient industries is emphasized with respect to the 45-industry analysis while the composition of the economy factor is larger as a result of using more detailed data.

Table 6.3 Decomposition of the change in GHG emissions between 1992 and 2006: impact of the disaggregation level (percentage)

	45 industries	*8 industries*
Emission intensity of energy use	–9.5	–7.1
Energy intensity of output	–11.2	–18.2
Structure of production	–4.5	0.1
Economic growth	30.8	30.7
Total change	5.6	5.6

Decomposition of acidification-related emissions

Absolute decoupling between the economic variables and the pollutants that contribute to acidification is ongoing; indeed, over the period 1992–2006 acidification-related emissions decreased by about 55.8 thousand tonnes (Table 6.1). This result was achieved thanks to the large contribution of the two technological components and to the structure of the production effect (their aggregate contribution is –73.1 per cent), whereas economic growth alone would have increased these emissions by 22.5 per cent (Figure 6.4).

Figure 6.5 shows that within this period the emissions intensity of the energy use effect causes a quite steady decrease in total emissions change (yearly average is about -4 per cent). Energy use intensity of output, on the other hand, shows that most of this total change occurred over the first half of the period (a 12.8 per cent reduction), and that in the second half of the period this component oscillates and its trend is less clear. With regard to the economic growth component most of the result at national level was achieved in the period 1994–2001.

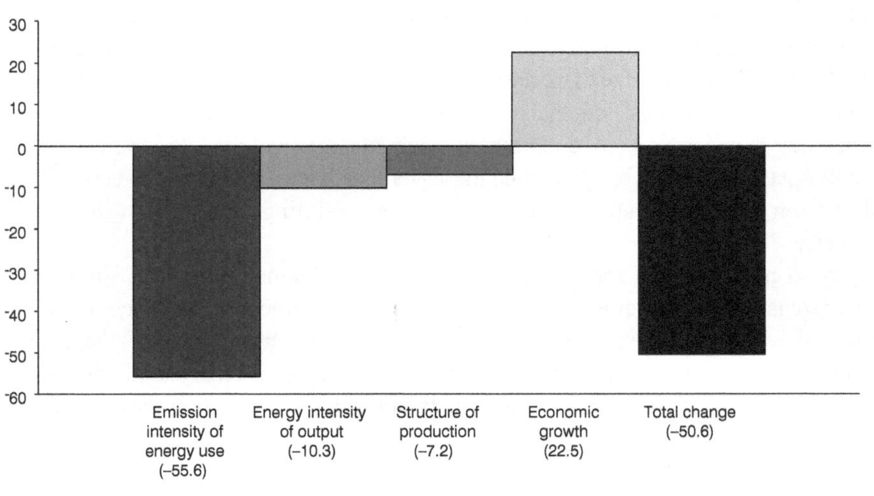

Figure 6.4 Changes between 1992 and 2006 in industry-related acidification emissions broken down by effect (percentage)

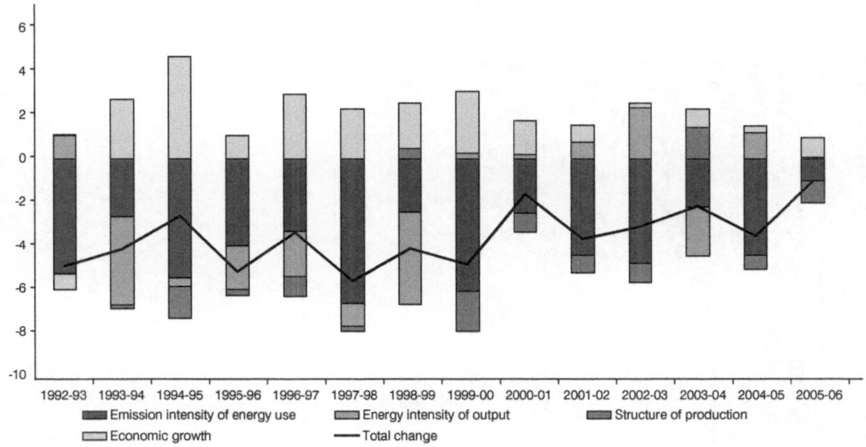

Figure 6.5 Year by year changes in industry-related acidification emissions by effect 1992–2006 (percentage of 1992 emissions)

By ranking industries according to their importance as acidification polluters the same industries as presented in Table 6.2 emerge, and again we find that the agricultural and the energy sectors are responsible for the greater part of total emissions (55.4 per cent), although they account for only 4.4 per cent of total output.[12] The transport sector is ranked next, contributing 12.3 per cent of Italy's production-related acidifying emissions.

The decomposition of acidification-related emissions is easier to understand than GHG emissions; indeed, for three out of four components, the performances reported in Table 6.4 have the same sign for all the most important economic activities. Only the structure of production shows positive values for some activities and negative ones for others. It should be noted that the two intensity factors contribute to an emissions decrease, though to varying extents. 'Agriculture, hunting and forestry' is characterized by a high value for the economic growth component, which, however, is largely offset by the negative contribution of the change in production structure (meaning that if this industry had grown in line with the whole economy, its emissions would have increased very significantly because of the weight of its acidifying emissions – 36 per cent of the total; relative to its weight of output, however, this did not occur).

The energy sector shows a noteworthy emissions intensity of the energy use factor which causes most of the total emissions change at both industry and national levels; considering that the energy intensity of output accounts for a small amelioration, we can say that the improvement in the other intensity component (–24 per cent) is due both to investments in better technologies (fuel being equal) and to the switch to a less polluting fuel (from coal and oil to natural gas). The acidification-related emissions in transport activities decreased between 1992 and 2006 due to technological factors; a remarkable improvement

Table 6.4 Decomposition of the change in acidification-related emissions of the most polluting industries between 1992 and 2006 (as percentage of the total amount of acidification-related emissions emitted by all industries in 1992)

Industries (NACE Rev. 1.1)		*Economic growth*	*Structure of production*	*Energy intensity of output*	*Emission intensity of energy use*	*Total change*
01–02	Agriculture, hunting and forestry	7.6	–6.8	–2.9	–2.1	–4.2
40–41	Electricity, gas and water supply	5.2	0.3	–1.5	–24.0	–20.0
60–63	*Transport*	2.6	1.3	–1.1	–6.4	–3.6
23	Manufacture of coke, refined petroleum products and nuclear fuel	1.8	–1.6	–0.4	–5.6	–5.8
26	Manufacture of other non-metallic mineral products	1.1	–0.3	–0.7	–1.2	–1.1
51–52	*Trade*	0.8	–0.1	–1.9	–2.2	–3.4
24	Manufacture of chemicals and chemical products	0.7	0.1	–0.2	–5.5	–4.9
27	Manufacture of basic metals	0.4	0.0	–0.5	–0.6	–0.6
17	Manufacture of textiles and textile products	0.3	–0.1	–0.2	–1.5	–1.6
90–91–93	Sewage and refuse disposal, sanitation and similar activities; Activities of membership organisations; other services	0.2	–0.1	0.5	–0.5	0.1

in the emissions intensity of energy use in particular, allowed a 3.6 per cent reduction in this industry.

The five manufacturing industries show substantially similar patterns of performance; they show improvements in the two intensity factors and small changes in the two economic components, and a significant improvement in emissions intensity of energy use in NACE 23 (refineries) and NACE 24 (chemicals).

In trade activities, both intensity components combine to bring about the –3.4 per cent contribution to total change; finally, the results for the NACE 90, 91 and 93 service activities do not show relevant contributions to total emissions change for any of their components.

Decomposition of tropospheric ozone precursor emissions

The reduction in tropospheric ozone precursors between 1992 and 2006 is some 1.9 million tonnes (Table 6.1). Figure 6.1 shows that the trends in acidification-related and tropospheric ozone precursor emissions are quite similar; Figure 6.6 shows that for these four effects the same changes are the same as in Figure 6.4. Hence, the economy-wide analysis for this theme, is not very different from that shown for the previous theme. In fact, we note a –51.4 per cent total change stemming from a –72 per cent decrease in emissions due to technological components and the composition of the economy factor, and a 20.6 per cent increase due to economic growth.

In Figure 6.7 the yearly analysis concerning all industries again shows the trend depicted in Figure 6.5 for acidification. However, different pollutants – and thus the different economic activities that give rise to them – are involved in the two themes; this is why it is important to analyse what happens, industry by industry.

Table 6.5 presents the results for the most relevant industries in terms of tropospheric ozone precursor emissions, for the period 1992–2006.[13] Again, the two technological components determine the reduction in total emissions: *Transport*, *Trade* and 'Other business activities' were able to take advantage of the transport policy introducing catalytic converters (Euro standards); the energy sector introduced more efficient technologies and less polluting fuels; fugitive emissions from fuels in the 'Sale, maintenance and repair of motor vehicles and motorcycles, and retail sale of automotive fuel' decreased based on more efficient technologies in retail fuel sales. The manufacturing industries, however, do not show similar improvements, probably due to the lack of appropriate economic incentives, including policy-determined ones, for fuel substitution and abatement technologies.

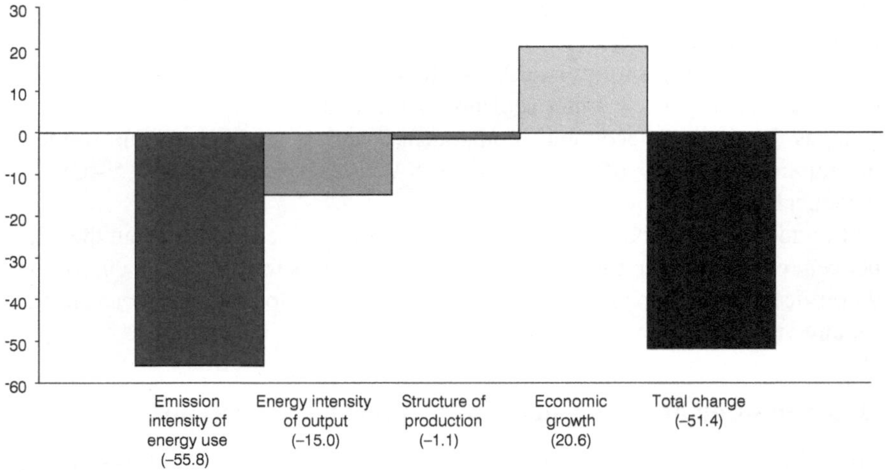

Figure 6.6 Changes between 1992 and 2006 in industry-related tropospheric ozone precursor emissions broken down by effect (percentage)

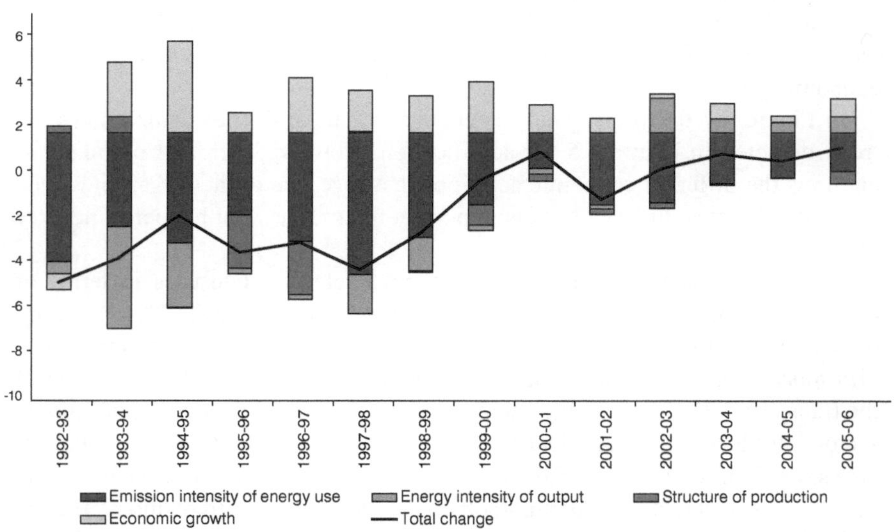

Figure 6.7 Year by year changes in industry-related tropospheric ozone precursors emissions by effect 1992–2006 (percentage of 1992 emissions)

Table 6.5 Decomposition of the change in tropospheric ozone precursor emissions of the most polluting industries between 1992 and 2006 (as percentage of the total amount of tropospheric ozone precursor emissions emitted by all industries in 1992)

Industries (NACE Rev. 1.1)		*Economic growth*	*Structure of production*	*Energy intensity of output*	*Emission intensity of energy use*	*Total change*
60–63	*'Transport'*	3.7	1.9	–2.6	–9.2	–6.2
40–41	Electricity, gas and water supply	2.4	0.2	–0.8	–11.1	–9.3
01–02	Agriculture, hunting and forestry	2.1	–1.8	–0.8	–1.4	–2.0
51–52	'Trade''	2.1	–0.2	–5.3	–8.4	–11.9
26	Manufacture of other non-metallic mineral products	1.2	–0.3	–0.8	–1.1	–1.0
23	Manufacture of coke, refined petroleum products and nuclear fuel	0.9	–0.8	–0.2	–1.4	–1.6
45	Construction	0.7	–0.4	0.3	–1.0	–0.4
27	Manufacture of basic metals	0.7	0.0	–0.7	–0.2	–0.3
24	Manufacture of chemicals and chemical products	0.6	0.0	0.0	–2.8	–2.2
28	Manufacture of fabricated metal products, except machinery and equipment	0.6	0.5	–0.7	–1.5	–1.2
74	Other business activities	0.5	0.3	–1.2	–2.3	–2.6
50	Sale, maintenance and repair of motor vehicles and motorcycles; retail sale of automotive fuel	0.5	0.1	–1.5	–1.3	–2.2

Conclusions and identification of further research needs

Decomposition analysis is a useful method for studying the determinants that affect environmental pressures over time and, in this respect, hybrid accounts provide an appropriate information base for correctly connecting economic, environmental and technological variables. The industry level analysis shows wide variations between different industry groups and the results reveal that individual industries do not always follow the general tendency (see some interesting cases of inverse patterns). The changes taking place in electricity, agriculture and transport activities, representing between 40 and 70 per cent of total emissions from production, greatly affect national decompositions with regard to all three themes. Other outstanding industries are: 'Manufacture of other non-metallic mineral products' for GHG; 'Manufacture of coke, refined petroleum products' for acidification; and *Trade* for tropospheric ozone precursors.

It should be noted that mathematically distinguishing the effects of the changes in determinants is not to imply that we can assume that the determinants are independent of each other. On the contrary, it should be remembered that the factors usually considered in this kind of analysis are often interconnected by functional links. For example, if one factor (e.g. emissions intensity/output unit) depends on the *efficiency* of a set of activities, and another factor expresses the *scale* at which those activities take place (such as output), it might well be that the former depends, at least in part, on the latter (i.e. there may be economies of scale). This is exactly what is hypothesised by the the so-called environmental Kuznets curve. To include in the analysis the results of quantitative studies of these links would be a useful step forward. In particular, a reduction of all the determinants to more basic factors could lead to the identification of truly independent causes of changes in emissions: on present evidence, all we can say is that emissions would have changed that much had the emissions intensity of output changed as it did but no economic growth had occurred. However, we know that this hypothesis is unrealistic, as emissions intensity changes in part for exactly the same reasons as economic growth occurs, that is primarily because of technological progress. We can apply the same reasoning to the two factors of economic growth and change in the composition of the economy, since the two are surely linked.

A different, though not necessarily alternative, strategy for progressing the research on the causes of emissions changes would be a further splitting of the determinants considered. In particular, the energy input from fuels per output unit determinant could usefully be broken down into the product of an energy input from fuel on the total energy input factor and a total energy input per output unit factor; this splitting would allow the identification and quantification of the contribution of the shift to renewable energy sources that is analysed here, as if it were a reduction in the energy input overall. Also the effects of the international delocalisation of production can be studied within this framework by analysing in greater detail the energy intensity component. Similarly a further split would allow us to distinguish in the component emissions intensity of energy use, the importance of the fuel mix in relation to other technological improvements.

Further research is certainly needed to achieve a deeper understanding of the causes underlying the changes in the factors highlighted by the decomposition exercise presented here. Possible next steps might include sector-specific analyses of the innovations introduced and existing incentives (e.g. for fuel substitution and for technological improvements such as the installation of abatement devices); of subsidization and taxation as well as of other environmental and economic policies, etc. Decomposition results, such as those presented above, provide important inputs to these kinds of analyses, as they broaden our knowledge of the phenomena that need to be studied, understood and, possibly, steered by policy.

Notes

1 Aldo Femia authored the introduction and conclusions, and Renato Marra Campanale the part on analytical results.

2 See Ang and Zhang (2000) for a list of 124 studies of the decomposition technique by application area. This technique allows the changes to be broken down into one variable according to the contribution of the changes in its determinants. The decomposition approach in this analysis builds on the general extension of the Dietzenbacher and Los (1998) method suggested by Seibel (2003), which is equally well suited for use with the Input–Output model and with sector-level data. The method is based on simulating what would have happened if one component had changed and all others remained fixed.

3 NAMEA tables are published annually by the Italian National Institute of Statistics (Istat; http://www.istat.it/salastampa/comunicati/non_calendario/20090128_00/). The principle of identity of the activities to which the different data refer is illustrated easily by the example of transport: it would be incorrect to refer all the emissions from cars and trucks to road transport activity (NACE division 60), since many road vehicles are operated by households or, as internal ancillary activity, by economic units belonging to other NACE divisions. In the NAMEA, the emissions from these ancillary activities (i.e., in our example, from road vehicles operated by economic units whose products are not transport services) are attributed to the activities they serve and not to the road transport division. Moreover, the transport activity, as intended in the NAMEA, produces emissions from processes other than vehicles, e.g. from the heating of the premises of the firms operating in the sector.

4 The GHG theme includes other, less important, compounds, but GHG calculations include only CO_2, CH_4 and N_2O. Conversion factors to CO_2-equivalents are made using the Global Warming Potential (GWP): CO_2: 1; N_2O: 310; CH_4: 21.

5 Conversion factors to PAE: SO_2: 1/32; NO_x: 1/46; NH_3: 1/17.

6 Conversion factors to TOFP: NO_x: 1.22; NMVOC: 1; CO: 0.11; CH_4: 0.014.

7 If the economic variable shows positive growth, absolute decoupling is said to occur when the rate of growth of the environmental variable is zero or negative. Relative decoupling is said to occur when the rate of growth of the environmental variable is positive but less than the rate of growth of the economic variable (OECD 2002).

8 This period allows the highest feasible level of disaggregation, namely 45 industries.

9 Usually, the importance of this component decreases when output is replaced by value added.

10 Bracci (2006), Coro and Volpe (2006), Costa (2007), Costa and Ferri (2007), Istat (2008).

11 Since the economic growth driver has the same value for all industries, the volume effect is always positive and its measure only depends on the other factors, which differ industry by industry. If an industry grows more (less) than average its share in the economy increases (decreases) and the structure effect is positive (negative).

12 Also, in this case, the ranking would be different had the industries been ranked according to the change in emissions rather than according to average share of their emissions. And, in this case, the set of the most important activities would be substantially unchanged.

13 The list of activities is built in a slightly different way from the two previous themes, i.e. it includes not only the most relevant groups of activities in terms of emissions shares, but also two additional activities which are in the 'top ten' for the contribution they provide to the overall *change* in these kinds of emissions. These are: 'Other business activities', and 'Sale, maintenance and repair of motor vehicles and motorcycles and retail sale of automotive fuel'; which are ranked respectively 13th and 15th among the list of industries with the highest average emissions shares. *Trade* and 'Other business activities' have relevant transport ancillary activities.

References

Ang, B.W. and Zhang, F.Q. (2000) 'A survey of index decomposition analysis in energy and environmental studies', *Energy,* 25: 1149–76.

Bracci, L. (2006) 'Una misura della delocalizzazione internazionale', in ICE (ed.) *L'Italia nell'economia internazionale*, Rapporto ICE 2005–2006, Rome: ICE, pp. 231–8.

Corò, G. and Volpe, M. (2006) 'Apertura internazionale della produzione nei distretti italiani', in Corò G., Tattara G. and Volpe M. (eds.), *Andarsene per continuare a crescere. La delocalizzazione internazionale come strategia competitive*, Rome: Carocci, pp. 21–53.

Costa, S. (2007) 'La delocalizzazione nel settore manifatturiero italiano: primi risultati dall'inchiesta Isae 2007', in ICE (ed.) *L'Italia nell'economia Internazionale*, Rapporto ICE 2005–2006, Rome: ICE, pp. 401–6.

Costa, S. and Ferri, G. (2007) 'The Determinants and Employment Effects of International Outsourcing: The Case of Italy', Dipartimento di Scienze Economiche, Università di Bari, *Working Paper 16*, Bari: University of Bari.

Dietzenbacher, E. and Los, B. (1998) 'Structural decomposition techniques: Sense and sensitivity', *Economic Systems Research*, 10: 307–23.

Istat (2008) 'Internazionalizzazione delle medie e grandi imprese (International sourcing)', http://www.istat.it/salastampa/comunicati/non_calendario/20080318_00/, Rome: Istat, accessed 18 March 2008.

OECD (2002) *Indicators to Measure Decoupling of Environmental Pressure from Economic Growth,* SG/SD(2002)1/FINAL, Paris: OECD.

Seibel, S. (2003) 'Decomposition Analysis of Carbon Dioxide Emission Changes in Germany – Conceptual Framework and Empirical Results', *Eurostat Working Papers and Studies*, ISBN 92-894-5167-X, Luxemburg: Eurostat.

7 Regional and sector environmental efficiency

Empirical evidence using the Italian RAMEA

Massimiliano Mazzanti and Anna Montini

Introduction

This paper develops empirical analyses using the NAMEA (National Accounting Matrix for Environmental Accounts) with reference to the Lazio region, Italy, where Rome is located. The data in our analysis are for 2000, the only year that both regional and national level data are available (national level data only are available for the period 1990–2003). By comparing regional and national environmental sector intensities, we aim to demonstrate the utility of NAMEA and shift share analysis for environmental and industrial policy making.

In referring to a regional framework, the analysis is very significant since it allows the investigation to focus on structural and idiosyncratic features compared with national averages, providing food for thought for regional development policy making, which is the key stone of economic development. It enables economic policies to be differentiated by regions, on the basis of the observed heterogeneity in economic-environmental relationships.

We are aware of some rare examples at the international level, of regional analyses, and also a few national level studies, including the work carried out by Wuppertal Institute on environmental input–output methodologies (Nansai *et al.* 2007; Suh 2005; Huppes *et al.* 2005) including NAMEA-like data that have covered mainly emission but also waste and materials (Nakamura 1999; Moll *et al.* 1999), and good-quality Spanish data (Roca and Serrano 2007a,b).

For an overview of the recent development of regional NAMEA (RAMEA) in Italy see the institutional site www.arpa.emr.it/ramea. Stauvermann (2007) and Goralzcyck and Stauvermann (2008) present some comparative environmental performances of a RAMEA EU project.

At the international level there are some academic works, such as Ike (1999), Vaze (1999), and Keuning *et al.* (1999), which present and discuss some country-specific NAMEA experiences, within the perspective of structural change analysis. Steenge (1999) provides a policy-oriented analysis with reference to the possible policy implications of NAMEA. There are a very few studies based on a proper environmental economics oriented perspective, for example Mazzanti *et al.* (2008), who exploit panel data for Italy to assess environmental Kuznets curves dynamics over 1990–2001.

RAMEA emerging frameworks

Within the recent studies that exploit NAMEA data, it is highly relevant to place attention and some more comments on Stauvermann (2007), who presents a Dutch pilot study on a regional RAMEA. This work shares many elements with the analysis we present here and, moreover, is highly complementary in suggesting future research directions. In fact, this work and research project is highly relevant and complementary to our present analysis, in the aim to bring together different European research experiences on NAMEA and RAMEA, in light of establishing a future EU-based NAMEA. A standardization of different experiences that have so far developed more or less independently would be essential, at least for major countries, in order to allow comparisons of various evidences and performances in the income–environment indicators of *sustainable production and consumption* (Watson and Moll 2008), where trade issues play a major role. As also argued by Stauvermann (2007), the integration of trade flows in NAMEA by linking different country accounting systems is a valuable research effort for the future. At the moment, though methodological issues are clear, empirical limitations given by data availability limits such research or circumscribes it to national case studies for countries which have sufficient data, such as the UK, Norway and Denmark (Harris 2001; Muradian *et al.* 2002). Still in this case, data constraints make necessary strong assumptions regarding the emission intensity of trading partners, in the absence of real NAMEA data for other countries. In other cases (Mazzanti *et al.* 2008), the integration of NAMEA data could be limited to the inclusion of trade openness indicators among the drivers of environmental efficiency, in order to infer whether the forces behind the pollution haven hypothesis or alternatively those driving production specialization in energy/environment intense sector prevail. These analyses refer to the old debate around the Leontief paradox which has witnessed recent interesting insights even in the environmental realm (Dietzenbacher and Mukhopadhay 2006).

In addition, such series of integrated NAMEA and EU-level evidences may complement the EKC-oriented evidence, which stems from econometric analysis of the dynamic income–environment relationship, which is robust in assessing dynamic facts and environmental pressures drivers, but often lacks a comprehensive investigation of structural factors behind the increase or decrease of environmental efficiencies.

The Dutch RAMEA also connects to the various RAMEA experiences that ISTAT and Italian regions have been developing over the recent years, with the objective of reaching the establishment of a full Italian RAMEA covering all or most of the 20 regions, which will allow more robust shift share analysis in the future and eventually, if a sufficient, at least two years (available for Italy in 2010–11), panel dataset is created, even structural decomposition analysis robustly rooted in a NAMEA matrix environment. For a survey on recent works on environmental decomposition analyses within which the shift share approach is embedded we refer to the extended version of this work (Mazzanti and Montini 2009).

Stauvermann focuses on the environmental and economic aspects of a region, but proposing future extensions on social dimensions that could be added to set a RAMEA.

Among the many complementary and interlinks between the works, Stauvermann (2007: 13) notes that 'the aggregate environmental damage of a country or of a region does not only depend on the country's size and development stage but also on its structure of economic activities'. This is a value added that NAMEA, and with increasing detail RAMEA, possesses, with respect to decoupling analysis based on the EKC framework that often lacks an assessment of structural factors of change as an additional explanatory factor to the core income–environment relationship.

We highlight that the present chapter, which exploits official ISTAT data for Lazio recovered from a sample of regional local emission sources, does not face the problem that Stauvermann acknowledges, that is the fact that, if value added is available at regional and sector level, environmental information on emissions exists only at the national level, so that we must estimate these data. Based on the assumption of country homogeneity, regional environmental data are derived directly by distributing the national environmental data in the same way as the distribution of environmental data to economic data. Though limited, the reliability of the method is testable from a statistical perspective. A real regional emission dataset, though possibly exposed to other estimation biases, is the priority, if feasible. This is a key issue in the establishment of compatible and robust RAMEA data. While the Netherlands faced this problem, to our knowledge the Lazio RAMEA was constructed by ISTAT using a bottom-up approach from regional inventory sources provided by ISPRA (Italian environment agency).[1] Regional inventory data were also used by Tuscany; instead, Emilia Romagna, though possessing a regional inventory, faced problems that regarded the integration of NACE sector with emission SNAP codes that constitute the origin of sector emission data.

It is interesting to comment on and compare the set of ecological-economic indicators Stauvermann proposes, as an alternative, or better a first step embedded in a proper shift share analysis that compares regional and national data in this case. First, sector environmental impact indicators and environmental efficiencies are compared by means of a normalization of the regional average, to highlight what sectors are more or less eco-efficient than the regional average. This analysis is first carried out just on emission-ecological factors and then by using economic-environmental indicators (emission/value added ratio) as we exploit in our paper. In all cases the comparison is merely between the regional average of the indicator and the sector-specific values, or eventually the regional eco-efficiency and the national eco-efficiency per sector. Finally, a synthetic index can be set up by relating the emission share of a sector, and its economic share, to the respective regional average shares. This relative indicator, which captures the extent to which the sector is contributing in emissions more or less with respect to its economic impact (less, if for example the emission share is lower than that of value added, thus the index is lower than unity), leads to conceptual frameworks which have stricter connection with shift share (this may be an embryonic component of it) and delinking / environmental efficiency and productivity analyses.

In addition, what lacks and instead is the core value of our chapter is the analysis of drivers when decomposing the emission/value added index in its potential sources.

The Dutch experiment, in comparison to our current one, covers three years, 2001–2003. This prevents us from carrying out a decomposition analysis that will be possible in the next evolution of regional NAMEA in Italy. On such a basis, a dynamic shift share analysis is attempted for 2001 and 2003 (following a static analysis of 2002) by identifying the three components of national share, regional shift, industry mix, in order to account for regional ecological competitiveness. We would remark both ecological *and* economic competitiveness, given that environmental efficiency calculated in terms of emission/value added is a real component of economic performance, driving both private[2] and public (non-market) benefits.

This shift share methodology emerged in the 1960s as a tool for analysing the indicators of regional productivity and employment (Dunn 1960). It has since been applied to other issues, such as international trade, and more recently tourism economics but, to our knowledge, it has never been used for environmental economic analysis. The specific methodology we use here was introduced by Esteban (2000, 1972).

Objectives of the study and the empirical model

The first empirical objective of this chapter is to measure the role of the regional productive structure in explaining the emissions efficiency gap between Lazio and Italy, using shift share analysis. We obtained a figure for the role of the specific emissions efficiency for the macro sectors defined. The decision to use shift share analysis was based on the need to determine the effects and factors that synthetically explain the relative efficiency/inefficiency of the regional system compared with the (national) average. Our aim is to examine and test whether the gap between the region under consideration and the benchmark average depends on an overall higher/lower productivity differential for all sectors, and/or on a higher/lower regional specialization in sectors with higher/ lower productivity.

In our analysis, primary attention is on the intensity of emissions, in other words on the indicators of emissions per value added, at sector level, given that this variable provides insights into the efficiency of productive sectors, which is a very useful information for the formulation of actions to support environmental innovation at sector level.

More specifically, we develop an analysis of the relative environmental efficiency of the Lazio economic system with respect to the national average, referring to a vector of ten pollutants, which encompass greenhouse gases, regional pollutants and local pollutants, and to the economic sector included and specified by NAMEA.

Our starting point is the aggregate indicator of emissions intensity, represented by 'total emissions on value added', defined as E/VA for Italy (the benchmark), and as E_1/VA_1 for Lazio. This indicator is decomposed as the sum of $(E^s/VA^s)*(VA^s/VA)$, where VA^s/VA is the share of sector value added on total value added, for all sectors s, with the value of s defined from one to j ($j = 24$, that is the number of NACE sectors included in the regional NAMEA).

For clarity, we redefined the index of emissions intensity as X for the national average ($X = E/VA$), as X_1 for Lazio ($X_1 = E_1/VA_1$), and as X^s for each sector (for Lazio $X^s_1 = E^s_1/VA^s_1$, for Italy $X^s = E^s/VA^s$). We then defined the share of sector value added as $P^s = VA^s/VA$ for Italy and $P^s_1 = VA^s_1/VA_1$, for Lazio.

In other words:

$$X = \sum_s P^s X^s \quad (7.1)$$

$$X_l = \sum_s P^s{}_l X^s{}_l \quad (7.2)$$

On that basis we can easily identify three effects, as prescribed by the *shift share decomposition*. These three effects explain the gaps in terms of aggregate emissions efficiency, between Lazio and Italy.

The first effect (structural or *industry mix*) is given by:

$$m_l = \sum_s (P^s{}_l - P^s) X^s \quad (7.3)$$

m_l assumes a positive (negative) value if the region is specialised ($P_l^s - P^s > 0$) in sectors associated with lower (higher) environmental efficiency, given that the gap in value added sector shares is multiplied by the value X of the national average (as if the region was characterized by average national efficiency). The factor m_l assumes lower values if the region is specialised in (on average) more efficient sectors.

The second factor, defined as the differential or efficiency, is:

$$p_l = \sum_s P^s (X_l^s - X^s) \quad (7.4)$$

p_l assumes a positive (negative) value if the region is less (more) efficient in terms of emissions (the *shift* between regional and national efficiency), under the assumption that value added sector shares were the same for the region, and for Italy ($P_l^s - P^s = 0$).

Finally, the effect of covariance between these two equations, or the allocative component, is given by:

$$a_l = \sum_s (X_l^s - X^s)(P_l^s - P^s) \quad (7.5)$$

The a_l factor is positive (negative) if the region is not (is) specialised, relative to the national benchmark, in sectors characterized by higher (lower) emissions intensity.

It assumes a minimum value, in our case, if the region is specialised in sectors for which it presents the highest comparative advantage (low intensity of emissions) thus, the covariance factor is between m_l and p_l.

The total difference in emissions intensity between the region and Italy, for each pollutant, is the sum of the aforementioned factors:

$$X_l - X = m_l + p_l + a_l \tag{7.6}$$

This decomposition into three factors allows a quantitative and synthetic measure of the underlying reasons for the differences in emissions intensity. It allows assessment of the aggregate differentials we observe. For example, it may be that a higher value for the regional emissions intensity depends only on productive structural motivations, on which environmental and energy policy can have no direct impact. Such policy would be more effective at altering the dynamics if the gap were due relatively more to a specific sector inefficiency, attributable to technological factors or/and inadequate organizational and regulatory frameworks.

Empirical evidence

First, we look at the evidence for the aggregate efficiency indicator $(X_1 - X)$.[3] It is clear that Lazio emerges as being relatively more efficient for all the pollutants and emissions considered (Table 7.2). The sector decomposition also shows the extent to which the comparative advantage in efficiency is derived from services (G–P branches), and some manufacturing branches (DE, DF-DG, DJ, and the aggregate DK-DL-DM), which do not show a gap that does disfavour to the region for any emissions.

This empirical information is not sufficient, however, to identify the main drivers of the efficiency differential, or to provide major implications for policy.

Therefore, we next analysed (Table 7.3) the factors and components (*m, p* and *a)* that contribute to explaining the $(X_1 - X)$ differential. We note that, in eight out of ten cases, including greenhouse gases and the main regional acid rain pollutants and local pollutants, the primary finding from the shift share analysis is the efficiency factor (*p*), which favours Lazio. Its relevance is associated to a weight that is often more than 50 per cent of the difference we observe between the region and Italy.

Finally, some comments on the results of the shift share analysis on the aggregates of manufacturing sectors (D), services (G–P) and other industrial sectors (C, E, F). Note that this investigation does not affect the regional comparative advantage for *all* NAMEA emissions.

The differences $(X_1 - X)$ are negative for all macro sectors and all pollutants. Also, we can verify whether this higher efficiency is higher or lower in the three macro sectors with respect to the average benchmark related to the region–Italy comparison. In other words, the analysis by macro aggregates shows the extent to which they contribute to the average advantage of the region.

Table 7.1 Value added by productive branches. Lazio and Italy – year 2000 (shares)

Productive branches (ATECO 2001)		*Value added shares*	
Title	*Nace Code*	*Lazio*	*Italy*
Agriculture, hunting and forestry	A	0.016	0.030
Fishing	B	0.000	0.001
Mining and quarrying	C	0.001	0.004
Manufacture of food products, beverages and tobacco	DA	0.011	0.020
Manufacture of textiles and textile products	DB	0.005	0.006
Manufacture of leather and leather products	DC	0.000	0.023
Manufacture of wood and wood products, Manufacture of rubber and plastic products, Manufacturing n.e.c.	DD-DH-DN	0.010	0.026
Manufacture of pulp, paper and paper products	DE	0.016	0.015
Manufacture of coke, refined petroleum products and nuclear fuel, Manufacture of chemicals, chemical products and man-made fibres	DF-DG	0.025	0.020
Manufacture of other non-metallic mineral products	DI	0.008	0.014
Manufacture of basic metals and fabricated metal	DJ	0.007	0.031
Manufacture of machinery and equipment n.e.c., Manufacture of electrical and optical equipment, Manufacture of transport equipment	DK-DL-DM	0.032	0.059
Electricity, gas and water supply	E	0.027	0.022
Construction	F	0.040	0.050
Wholesale and retail trade; repair of motor vehicles, motorcycles and personal and household goods	G	0.121	0.138
Hotels and restaurants	H	0.029	0.035
Transport, storage and communication	I	0.114	0.078
Financial intermediation	J	0.091	0.066
Real estate, renting and business activities	K	0.197	0.181
Public administration and defence; compulsory social security	L	0.085	0.051
Education	M	0.046	0.044
Health and social work	N	0.046	0.044
Other community, social and personal service activities	O	0.054	0.036
Household related activities	P	0.016	0.008
Total		1.000	1.000

Table 7.2 Emission intensities. Lazio and Italy – Year 2000 (emission tonnes per M€ of value added)

NAMEA emissions/pollutants	*Lazio*	*Italy*
CH_4	1.148	1.769
CO	0.874	1.793
CO_2	221.860	381.072
N_2O	0.054	0.130
NH_3	0.179	0.435
NMVOC	0.470	0.750
NO_x	0.763	1.106
Pb	0.000211	0.000329
PM_{10}	0.069	0.165
SO_x	0.260	0.779

This comparative assessment, which was made by comparing results in Tables 7.3 and 7.4 (the table showing the actual comparison is omitted here, but is available upon request), indicates quite clearly that there are very few cases where the gap favouring the region in the overall analysis of the economic system, emerges as higher for manufacturing and services: four pollutants for manufacturing (CO, N_2O, NH_3 and, with minor emphasis, CH_4), and one case for services (PM_{10}).

Services, in comparative terms, is the aggregate sector that is less efficient than the regional average, although we would emphasise that sectors G–P are always less intensive for emissions with respect to the national averages.

Thus, it can be said that the Lazio region's environmental comparative advantage is mainly driven by other industrial sectors (extraction of materials, production and distribution of energy, construction). As before, we observe that the main driver explaining the differential (*p*) is related to sector efficiency. We note the heterogeneity across macro sectors: factor (*p*) in six cases is the main driver of manufacturing, while for services and other industries it is the main driver respectively in nine and ten cases.

Conclusions and policy insights

Our shift share analysis aims to offer a relative performance between Lazio region and the Italian average, in terms of environmental efficiency as here defined. We summarize here some key critical outcomes and some policy considerations linked both to the current analysis and to the extension towards future research using updated NAMEA datasets.

We have shown that for all emission included in NAMEA the sum of the three shift share factors indicates that Lazio is comparatively more environmentally efficient than the regional average.

For most emissions, we may claim from our knowledge of the Italian framework (ENEA 2006) that the main source of this difference is a lower energy

Table 7.3 Shift share coefficients regarding the total economic system (all productive branches)

NAMEA emissions/ pollutants	X_l	X	$X_l - X$	*Difference %*	m	p	a	$m+p+a$	*Primary factor*	*Primary factor (%)**
CH_4	1.148	1.769	–0.621	–35%	–0.136	–0.471	–0.0130	–0.621	P	76%
CO	0.874	1.793	–0.919	–51%	–0.431	–0.770	0.283	–0.919	P	52%
CO_2	221.860	381.072	–159.212	–42%	26.429	–159.253	–26.388	–159.212	P	75%
N_2O	0.054	0.130	–0.076	–59%	–0.0272	–0.0428	–0.006	–0.076	P	56%
NH_3	0.179	0.435	–0.256	–59%	–0.186	–0.1105	0.041	–0.256	P	33%
NMVOC	0.470	0.750	–0.280	–37%	–0.162	0.0775	–0.194	–0.280	A	45%
NO_x	0.763	1.106	–0.343	–31%	0.0298	–0.297	–0.075	–0.343	P	74%
Pb	0.0002110	0.000329	–0.000118	–36%	–0.0002	–0.000040	0.0001	–0.000118	M	59%
PM_{10}	0.069	0.165	–0.097	–58%	–0.031	–0.0720	0.0072	–0.097	P	65%
SO_x	0.260	0.779	–0.519	–67%	0.118	–0.529	–0.108	–0.519	P	70%

Notes: * share calculated on the sum of components in absolute values.

Legend:

X_l = (total emissions Lazio/total value added Lazio)
X = (total emissions Italy/total value added Italy)
m = sum by sectors Σ $((VA^s_l/VA_l)–(VA^s/VA))*(E^s/VA^s)$
p = sum by sectors Σ $(VA^s/VA)*((E^s_l/VA^s_l)–(E^s/VA^s))$
a = sum by sectors Σ $((VA^s_l/VA_l)–(VA^s/VA))*((E^s_l/VA^s_l)–(E^s/VA^s))$
$X_l – X = m + p + a$

Table 7.4 Shift share coefficients regarding the analyses for Manufacturing (D), other industrial sectors (C,E,F) and Services (G–P)

NAMEA emissions/pollutants	X_I	X	X_I–X	*Difference %*	*m*	*p*	*a*	*m+p+a*
Manufacturing								
CH_4	0.261	0.421	–0.160	–38%	0.154	–0.194	–0.120	–0.160
CO	0.541	2.883	–2.343	–81%	–1.190	–2.2618	1.109	–2.343
CO_2	426.282	469.605	–43.323	–9%	90.967	–104.519	–29.771	–43.323
N_2O	0.027	0.163	–0.136	–83%	0.1788	–0.136	–0.178	–0.136
NH_3	0.001	0.047	–0.0456	–97%	0.0567	–0.045	–0.056	–0.0456
NMVOC	1.836	1.974	–0.138	–7%	0.2039	0.621	–0.963	–0.138
NO_x	0.964	1.091	–0.128	–12%	0.089	–0.146	–0.070	–0.128
Pb	0.001	0.001	–0.000003	–0.3%	–0.0005	–0.0001	0.0006	–0.000003
PM_{10}	0.146	0.273	–0.127	–47%	–0.039	–0.132	0.0447	–0.127
SO_x	0.691	0.852	–0.161	–19%	0.329	–0.346	–0.144	–0.161
Non manufacturing industries (other industrial sectors)								
CH_4	2.850	3.739	–0.888	–24%	1.340	–1.645	–0.583	–0.888
CO	0.747	1.664	–0.917	–55%	0.454	–0.996	–0.374	–0.917
CO_2	1315.702	2529.417	–1213.714	–48%	930.408	–1556.852	–587.270	–1213.714
N_2O	0.057	0.102	–0.044	–44%	0.035	–0.057	–0.022	–0.044
NH_3	0.002	0.003	–0.0009	–31%	0.00075	–0.00137	–0.00035	–0.0009
NMVOC	0.929	1.329	–0.400	–30%	0.110	–0.423	–0.087	–0.400
NO_x	1.693	2.764	–1.071	–39%	0.831	–1.385	–0.517	–1.071
Pb	0.00009	0.00011	–0.00002	–22%	0.00003	–0.00004	–0.00001	–0.00002
PM_{10}	0.204	0.363	–0.159	–44%	0.094	–0.189	–0.063	–0.159
SO_x	2.009	6.576	–4.567	–69%	2.480	–5.115	–1.931	–4.567

Services								
CH_4	0.651	0.706	–0.055	–8%	0.1999	–0.1978	–0.0566	–0.0546
CO	0.697	0.936	–0.239	–26%	0.0619	–0.2585	–0.0427	–0.2392
CO_2	97.181	112.641	–15.460	–14%	8.895	–23.946	–0.408	–15.460
N_2O	0.010	0.013	–0.0035	–27%	0.0018	–0.0046	–0.0007	–0.0035
NH_3	0.008	0.011	–0.0033	–29%	0.0025	–0.0047	–0.0011	–0.0033
NMVOC	0.205	0.255	–0.0495	–19%	0.0065	–0.0395	–0.0165	–0.0495
NO_x	0.575	0.784	–0.209	–27%	0.0732	–0.2616	–0.0209	–0.2093
Pb	0.000094	0.000117	–0.00002	–19%	0.000002	–0.000022	–0.000002	–0.00002
PM_{10}	0.024	0.067	–0.0427	–64%	0.0044	–0.0442	–0.0029	–0.0427
SO_x	0.053	0.146	–0.0932	–64%	0.0326	–0.1021	–0.0237	–0.0932

consumption per capita and a lower energy intensity (electric energy) on GDP, compared with national averages. Lazio presented in 2003 a value of 99.7 tep/million€GDP, the third lowest (Italian average is 126, with Lombardy, the most industrialised and richest region, showing 121). Then, the electric intensity was around 201.9 MWh/million€GDP, the lowest in Italy (288.4 on average and 301 for Lombardy). Finally, both the energy and electric energy intensity of the manufacturing sector is the lowest in Italy.

Nevertheless, starting from the relative efficiency and having discussed some structural facts behind this evidence, if we analyse more in depth the specific components, we may note some elements of criticalities for the region. We synthesise those elements in two main issues: the role of energy intensity and the role of services.

First of all, if we consider the sector composition factor, it turns out not to favour Lazio for CO_2, SO_x and NO_x, main environmental problems at supranational level. That is to say, regarding these three environmental externalities the regional economic system is relatively not favourable. This may depend on the strong role and weight of regional production of electricity through fossil fuel sources, which compensates for the low energy intensity. The region is in fact highly dependent on oil as a source (59 per cent), with natural gas at 21 per cent. Renewable energy, including hydroelectric in which Italy presents a comparative advantage (66 per cent of total renewable energy in Italy is from hydroelectric power stations that are mainly located in the north), plays a very minor role. This may explain the not favourable position, mainly regarding future scenarios, if we consider GHG emissions trends.

This not favourable status quo should be considered a primary target for environmental policies aimed at integrating the region in the national movement towards the EU policy targets that have recently witnessed a long-run decision of setting by 2020 a 20 per cent cut in GHG and a minimum 20 per cent threshold for the renewable content of energy production. This is challenging for the region, given that innovation dynamics in services are on average lower, and EU policy does not directly target services with environmental regulations that may be innovation drivers. Then, the low performance in renewables means on the one hand that the region possesses strong incremental possibilities, but with specialisation that should be invented, given the almost total absence of hydropower and wind power generation sites.

Secondly, even if Lazio is relatively more specialised than Italy on average in services, it arises that services are relatively less efficient compared with the performances of other branches within the region, even if they still benefit the region in comparison to Italy.

Evidence on energy intensities could provide some explanations on structural facts. The 2003 services intensity is on an average level in Italy (18.6 tep/million€GDP), and the same holds for the electric energy performance. The relevant services orientation of the region, and of Rome in particular, is on the one hand a value for environmental performances (productive specialisation effect), but is on the other hand partially compensated by a relatively high (in any case above average) energy intensity of the sector. This reflects an important point, mostly for local (municipality of Rome) policy actions: the high energy intensity of transport systems, which also connects to a high ratio of cars per head. Using as above ENEA (2006) data, we note that the region presented in 2003 an intensity of 50.7 tep/million€GDP, one of the highest in Italy (33.4 for Lombardy). Environmental and transport policies together should find complementary actions to tackle the relative low performance of the transport sector and of household behaviour towards transport, especially in the critical *Roman hot-spot*.

Thus, regional environmental performances do not depend *strongly* on its bias towards services, which is historically typical of the region and of Rome, and on which the recent strong economic growth, compared with Italy, is mainly based.

For the majority of pollutants, in fact, it is the second factor of shift share, the differential/sector efficiency component, that dominates quantitatively on the first effect, that of sector composition.

Overall, then, the relative environmental efficiency with respect to Italy is primarily explained by an actual lower intensity in emissions per unit of value added, more than by favouring sector composition and specialisation. Thermo-electric production fuelled by oil, an energy intensity transport sector in its broad meaning, and a service sector which on average is not performing under the Italian average regarding emission intensity, are the negative elements to which national and especially local policy makers should look for tackling future challenges of emission reductions and environmental efficiency increases in a broader context. The critical Rome hot-spot is especially relevant for the transport and services issues highlighted here.

The shift share analysis disaggregated for the three macro sectors highlights additional interesting insights. Total efficiency differentials still favour the region across pollutants and for all sectors. Nevertheless, relatively to the gap observed in the aggregate analysis, we note that the ranking of macro sectors in contributing to the regional environmental performance is the following: other industrial sectors (C, E, F), manufacturing and services. Services do not present cases of emissions for which their efficiency is higher with respect to the average regional efficiency, compared with Italy. Within the region, and this is a counterintuitive result with respect to qualitative at first sight assessment, environmental performance is not *primarily* driven by the structurally strong weight of services, and the dynamic evolution that witnesses an increasing share of such sectors, that characterizes Lazio more than the Italian average situation. This evidence may have relevant implications for a region like the one investigated here, in which the city of Rome plays a crucial role for economic and environmental performances.

This chapter thus shows that even with a single regional NAMEA, in connection with the country average NAMEA, is possible to highlight a series of facts which are helpful for the understanding of structural basis of the income–environment relationship, with an attention to energy issues, and for the definition of future national and regional policies. Panel data would surely offer more food for such analysis, though we note that such structural differences are changing in the medium long run. A real value added to our analysis would be offered either by a very long panel which is for now unlikely to exist, or by an extension of cross-section analyses to more regions. A two-year 20-region panel will be available for Italy during 2010–11, allowing massive improvements in regional and environmental analyses of economy-environment interactions.

Notes

1 Bonazzi and Sansoni (2008) present a shift share analysis for Emilia Romagna and a methodological discussion of RAMEA accounting systems.
2 Bruvoll and Medin (2003), Mazzanti and Zoboli (2009).
3 Table 7.1 shows the variable P_1 for Lazio and P for Italy, which is the decomposition for value added by each productive branch. Table 7.2 shows the variables X_1 (Lazio) and X (Italy), which refer to emissions on value added, by each pollutant. These four variables are the basis of the shift share analysis following the approach described above.

References

Bonazzi, E. and Sansoni, M. (2008) ‘Valutazione della efficienza emissive dei gas serra nella regione Emilia Romagna: una analisi statistica shift share a supporto dei decisori pubblici’, *Valutazione ambientale*, 13: 18–25.

Bruvoll, A. and Medin, H. (2003) ‘Factors behind the environmental Kuznets curve. A decomposition of the changes in air pollution’, *Environmental and Resource Economics*, 24: 27–48.

Dietzenbacher, E. and Mukhopadhay, K. (2006) ‘An empirical examination of the pollution haven hypothesis for India: towards a green Leontief paradox?’, *Environmental & Resource Economics*, 36: 427–49.

Dunn, E.S. (1960) ‘A statistical and analytical technique for regional analysis, *Papers and Proceedings of the Regional Science Association*, 6: 97–112.

ENEA (2006), *Rapporto energia ed ambiente (Report on energy and the environment)*, Rome: ENEA.

Esteban, J. (1972), A reinterpretation of shift share analysis. *Regional Science and Urban Economics*, vol.2, pp. 249–61.

Esteban, J. (2000) ‘Regional convergence in Europe and the industry mix: a shift-share analysis, *Regional Science and Urban Economics*, 30: 353–64.

Goralzcyck, M. and Stauvermann, P.J. (2008) ‘The usefulness of hybrid accounting systems for environmental policy. Advice regarding sustainability’, paper presented at the 2008 International Input Output Society meeting on Input–Output & the Environment, July 9–11, Seville, Spain.

Harris, R. (2001) ‘Methods for estimating air emissions from the production of goods imported into the UK’, paper prepared for DG Regional Policy and Eurostat, Working paper 2/2001/B/5, Brussels: Eurostat, European Commission.

Huppes, G., De Koning, A., Suh, S., Heijungs, R., Van Oers, L., Nielsen, P. and Guinee, J.B. (2005) ‘Environmental impacts of consumption in the EU: high resolution input output tables with detailed environmental extensions,’ *Journal of Industrial Ecology*, 10: 129–46.

Ike, T. (1999) ‘A Japanese NAMEA’, *Structural Change and Economic Dynamics*, 10: 123–49.

Istat (2007), La NAMEA Italiana, anni 1990–2003, Contabilità Ambientale, Rome: Istat.

Keuning, S., van Dalen, J. and de Haan, M. (1999) 'The Netherlands' NAMEA; presentation, usage and future extensions', *Structural Change and Economic Dynamics*, 10: 15–37.

Mazzanti, M. and Montini, A. (2009) 'Regional and Sector Environmental Efficiency Empirical Evidence from Structural Shift-share Analysis of NAMEA data', FEEM Working paper 11, Milan: FEEM.

Mazzanti, M. and Zoboli, R. (2009) 'Environmental efficiency and labour productivity: trade-off or joint dynamics?', *Ecological Economics*, 68: 1182–94.

Mazzanti, M., Montini, A. and Zoboli, R. (2008) 'Economic dynamics, emission trends and the EKC hypothesis. New evidence using NAMEA data for Italy', *Economic Systems Research*, 20: 279–305.

Moll, S., Femia, A., Hinterberger, F. and Bringezu, S. (1999) 'An Input–Output Approach to Analyse the Total Material Requirement (TMR) of National Economies', in Kleijn, R., Bringezu, S., Fischer-Kowalski, M. and Palm, V. (eds.) *Ecologizing Societal Metabolism: Designing Scenarios for Sustainable Materials Management*, ConAccount workshop proceedings, 21 November 1998, Amsterdam, CML report 148. Centre of Environmental Science (CML), Leiden, 39–46.

Muradian, R., O'Connor, M. and Martinez-Alier, J. (2002) 'Embodied pollution in trade: estimating the environmental load displacement of industrialised countries', *Ecological Economics*, 41: 51–67.

Nakamura, S. (1999) 'An inter-industry approach to analyzing economic and environmental effects of the recycling of waste', *Ecological Economics*, 28: 133–45.

Nansai, K., Kagawa, S., Suh, S., Inaba, R. and Moriguchi, Y. (2007) 'Simple indicator to identify the environmental soundness of growth of consumption and technology: eco-velocity of consumption', *Environmental Science & Technology*, 41: 1465–72.

Roca, J. and Serrano, M. (2007a) 'Income growth and atmospheric pollution in Spain: an input output approach', *Ecological Economics*, 63: 230–42.

Roca, J. and Serrano, M. (2007b) 'Atmospheric pollution and consumption patterns in Spain: an input output approach', FEEM Working paper 62, Milan: FEEM.

Stauvermann, P.J. (2007) 'The regionalized NAMEA-type matrix: Methodology, application and interpretation - A Dutch pilot study', mimeo.

Steenge, A. (1999) 'Input-output theory and institutional aspects of environmentalpolicy', *Structural Change and Economic Dynamics,* 10: 161-76.

Suh, S. (2005) 'Developing a sectoral environmental database for input output analysis: the comprehensive environmental data archive of the US', *Economic System Research*, 17: 449–69.

Vaze, P. (1999) 'A NAMEA for the UK', *Structural Change and Economic Dynamics*, 10: 99–121.

Watson, D. and Moll, S. (2008) 'Environmental benefits and disadvantages of economic specialisation within global markets, and implications for SCP monitoring', Paper for the SCORE! conference, 10–11 March, Brussels.

8 Biofuels public support and technological specialization in the energy sector

Valeria Costantini, Francesco Crespi and Annalisa Zezza

Introduction

Recent years have witnessed dramatic volatility in oil prices and growing concern about the consequences of carbon emissions from fossil fuels. The Energy, Transport and Environment Indicators published by Eurostat (2007), show that in 2005 the transport sector was responsible for about 31 per cent of total energy consumption in the European Union (EU_{27} members) and 19 per cent of total Greenhouse Gas (GHG) emissions. In the ten years between 1995 and 2005 there was a 21 per cent increase in energy consumption for transport, mainly driven by consumption for road transport, which represents the largest share (82 per cent) of total transport consumption. Since crude oil accounts for 96 per cent of total energy consumption in the road sector, the 17 per cent increase in GHG emissions in this sector in 1995–2005 from transport is fully explained. This ten–year trend shows that current European policies for sustainable energy are not adequate for this sector, which is highly dependent on fossil fuels, and that more effort is required on alternative energy sources.

There is growing interest in biofuels based on agricultural crops, worldwide, because they represent an inexhaustible energy source and could have a positive impact on reducing carbon dioxide (CO_2) emissions. World biofuel production and consumption are characterized by pervasive and large subsidies, which has important implications for their environmental goals, in terms of their cost effectiveness. The huge increase in biofuel production is controversial in three major aspects: energy accounting; the conflict between energy and food production; the potentially distorting effect on innovation and technological specialization. While the first two aspects have been analysed extensively by academicians and international organizations (OECD 2005; Tokgoz et al. 2007), the third has been less well examined.

As biofuels are just one of the alternative technologies currently available for addressing energy and environmental goals, the emphasis on biofuel support policies may be skewed in terms of technical progress in renewables and energy–saving technologies. This chapter investigates whether public support for the biofuel sector is producing distortions in energy sector technological specialization by diverting public and private investments from other renewable energies and energy–saving technologies.

We use a gravity equation model based on bilateral export flows of technologies for production and consumption of renewable energies (wind, solar and photovoltaic, fuel cells) and energy–saving technologies. The model used is similar to that applied in other empirical studies that focus on the effects of environmental regulation on trade flows, but our application has two major advantages. It uses data for many countries and many years, whereas many empirical studies of innovation and adoption of environmental technologies focus on a single country, or on rather generally defined environmental policies. Moreover, by using a gravity equation, it enables investigation of the roles of distinct demand and supply policies supporting biofuels, on the export flows of technologies for renewable energy and energy saving. As export flows can be considered a measure of the strength of the competition at the international level (based on comparative advantage), use of the gravity model helps to reveal whether support for biofuels is depressing the competitiveness of new energy technologies. If the public policy to support biofuels is diverting investments and reducing the competitiveness of energy–saving and renewable–energy technologies, this would imply a conflict among policy actions.

This problem could be particularly relevant in Europe, where the European Commission Energy Climate Action Plan or the '20–20–20 strategy', combines several policy actions, such as a binding target for all EU member states of 20 per cent of energy consumption coming from renewables, and a minimum 10 per cent market share for biofuels by 2020. Also, the European Commission (EC) has declared a further effort – to improve energy efficiency in order to reduce energy consumption by 20 per cent, by 2020.

The Action Plan states that the EU goal of saving 20 per cent in energy consumption by 2020 through energy efficiency is a crucial part of Europe's energy and climate policy, because it constitutes an important means of reducing CO_2 emissions. In our view, this is a clear example of a set of multiple policies setting conflicting goals and providing conflicting market signals to private investors.

Several alternative policy variables for public support to biofuels, such as separation of fuel mandates, and reductions in taxes and tariffs on import flows based on data on bioethanol and biodiesel, have been tested to underline the impact of policy on the international competitiveness of the technologies required to achieve the EU's ambitious energy and environmental policy targets.

Overview of biofuel sector: production, consumption and policy support

Due to the high dependence on energy and the need to reduce GHG emissions, biofuels for transport (e.g. ethanol and biodiesel) are attracting interest in many countries. Bioenergy is seen by many to play a key role in the short run in reducing table GHG emissions; and biofuels are the only suitable substitute for fossil fuels in the transportation sector. Transport consumes 30 per cent of global energy, 99 per cent of which is supplied by petroleum, and is expected to account for about half of the total projected increase in global oil use in 2003–2030 (IEA 2007). Global production of biofuels amounted to 9.8 EJ (exajoules) in 2005, about 1 per cent of total fuel consumption in transportation, with production increasing at high rates with projected market shares of around 13 per cent in 2050 (IEA 2007). This is based on response to market forces and is supported by government policy.

The major ethanol producers are the United States (US) with 24.6 billion litres in 2007, followed by Brazil with 19 billion litres: these two countries account for more than 87 per cent of world supply. Ethanol production is based on corn in the US and sugar cane in Brazil. EU ethanol production is very limited (2.1 billion litres in 2007) but is growing and represents 18.5 per cent of EU biofuel production. The main member state producers are Spain, France, Germany, Sweden, plus Poland – one of the accession countries. The main feedstocks for EU ethanol production are cereals, corn and sugar beet. The EU is a leader in production of biodiesel, obtained mainly from rapeseed, sunflowers and soybean. EU biodiesel production amounted to 5.7 millions tons in 2007, half of which came from Germany. Biodiesel production is increasing in the US.

Biofuel production costs vary significantly across the main producing countries. Brazil has the highest competitive advantage for ethanol and is the only producer, based on the current state of technology, that can compete with fossil fuels. All other producing countries have to adopt some form of policy intervention.

Policy instruments for biofuels belong to the spheres of energy policy, environmental policy, agricultural policy and fiscal policy, depending on different interests in biofuel production and consumption. Policy instruments cover a fairly large set of support and regulatory measures that can be adopted at national, regional or local level. They fall into three main groups: measures that impact mainly on supply; measures that impact mainly on demand; measures that impact on technology and market developments.

Supply side policy measures currently provide most of the support for biofuels (GSI 2007), and are based mainly on tax exemptions or fuel tax rebates for gasoline and diesel, or volumetric tax credits, and border protection via tariffs. Other forms of supply side intervention comprise support for feedstock production and research and development (R&D) efforts. Comparison of world policy incentives for ethanol production is presented in 8.1.

Most countries producing biofuels apply a most–favoured nation (MFN) tariff that adds at least 20 per cent to the cost of imported ethanol. Tariffs are lower in the case of biodiesel. The primary objective of border protection is to limit the benefits of direct support through fiscal policy to domestic producers. Various exceptions to MFN tariffs and tariff–rate quotas apply to countries involved in free trade agreements.

8.2 shows that tariff regimes vary by product and by country in determining the allocation of comparative advantage at world level.

In addition to border protection, most countries support domestic production of biofuels through favourable fiscal regimes that reduce the cost differentials with

Table 8.1 Country comparison of support to ethanol production (cent $/l)

Country	*Production incentive*	*Excise reduction*	*Import tariff*	*Reduced tariff*
Australia	–	28.9	28.9	None
Brazil	–	30	–	MERCOSUR
Canada	Up to 16.4	15.1	4.3	NAFTA, CAFTA, CILE
EU	–	Up to 70.9	24.1	EFTA, GSP (excluding Brazil)
Switzerland	–	57.8	27.7	EFTA, GSP(excluding Brazil)
USA	13.5 + state incentive	Up to 8.4	2.2% + 14.3	NAFTA, CBI

Source: Steenblik (2005a).

Table 8.2 Applied tariffs on biofuels in representative countries as 1/1/2006

	Ethanol[1]		*Vegetable oils*[2]	
Country	*Applied MFN*[3]	*Imports value (current US$ 000)*	*Applied MFN*[3]	*Import value (current US$ 000)*
Australia	5.0	2,771	4.1	16,544
Canada	13.1	20,398	1.0	7,906
Japan	11.1	212,566	6.6	71,363
Switzerland	43.4	18,514	70.0	15,213
United States	13.5	345,708	1.8	33,884
European Union	54.7	197,705	2.1	1,831,554

Notes:

(1) Tariffs on ethanol corresponds to the Applied MFN (AHS in UNCTAD–Trains) for the code 220710 in the Harmonized Standard classification (HS 1996)

(2) Tariffs for vegetable oils are computed as an average of the tariffs corresponding to the codes 120500, 150710, 151110, 151211, 151410 (HS 1996) weighted by the relative imports values.

(3) Values expressed as ad valorem equivalent (%), calculated as a weighted average of tariffs. The values represent the real tariff applied to international imports (AHS in Trains database) and not the declared Most–favored nation tariff (MFN).

Source: UNCTAD–TRAINS database.

gasoline or diesel. In the US, reduced taxes for ethanol were introduced in 1978. In 2007, the Volumetric Ethanol Excise Tax Credit provided a fixed tax credit of $0.51 per gallon of ethanol blended with motor gasoline (and $1.00 per gallon for biodiesel). The level of exemption does not adjust to changes in oil prices or additional state exemptions (8.3).

The EU does not have a community level fiscal regime, but via the biofuel directive (EU 2003) authorizes member states to grant fuel tax reductions within certain limits. Tax rebates range from €0.10 to €0.65 per litre with an average of around €0.30 per litre, and they can be applied separately to bioethanol or biodiesel, or to generally defined biofuels. Some countries, such as France and

Table 8.3 Value of excise tax reduction at 1/1/2007 (€/lt)

Country	*Ethanol*	*Biodiesel*
Australia	0.23	0.23
Brazil	0.108	0.08
Canada (Federal)	0.066	0.264
EU		
Austria	0.015	0.028
Belgium	0.59	0.367
Check Republic		0.292
Denmark	0.03	0.03
France	0.37	0.37
Germany	0.65	0.47
Hungaria	0.10	0.34
Ireland	0.368	0.368
Italy	0.26	0.413
Lituania	0.25	0.25
Spain	0.42	0.29
Sweden	0.15	0.18
UK	0.32	0.32
Switzerland	0.45	0.47
USA	0.104	0.10–0.20

Source: GSI (2007).

Italy, have adopted a production quota system where tax rebates apply up to a given amount of production.

Biofuels are based on agricultural crops and crop residues, whose costs of production range from 50 per cent of the total cost of production in the case of ethanol, up to 90 per cent for biodiesel. Thus, agricultural and agricultural products trade policies have little impact on biofuel economics. In the US, price support coupled with deficiency payments for agricultural products have substantially reduced the cost of biofuel feedstocks. Plus, in the US, up to 2006, specific direct support for biofuels was provided for in the Bioenergy Program. In the EU the biofuel sector receives support in the form of permission to grow biofuel crops on set–aside land and through the granting of direct area payment of €45 per ha to energy crops grown on other land.

Policies to increase demand for biofuels by substitution for fossil fuels include regulatory measures such as targets and mandatory requirements. While some of these do not discriminate among forms of biofuels, others specifically target ethanol and biodiesel.

The most significant mandates, mainly related to potential demand for biofuels, apply to the US and the EU although they are also in force in Brazil, China, India, South Africa and other countries. These kinds of measures are usually aimed at the medium term and are generally complemented by other measures designed to develop the biofuel market.

The US mandate was established within the Energy Bill in 2005 and is known as the renewable–fuel standard. It requires minimum consumption of biofuels from 11.9 million tons in 2006 to 22.1 million tons in 2012. The 2012 target level was almost achieved in 2007.

In the EU the EC Directive 2003/30 on biofuels fixed indicative targets of up to 2 per cent for 2005 and 5.75 per cent for 2010. The actual average blend rate was below 1 per cent in 2005, but nine member states have established mandatory blending requirements (see Table 8.4).

Table 8.4 Fuel target or mandate by country

Country	*Fuel target or mandate*	
	Type	*Quantity or blending share*
Australia	T	350 million liters by 2010
Canada	M	5% by 2010(ethanol)
		2% by 2012 (biodiesel)
EU	T	2% by 2005; 5.75% by 2010; 10% by 2020
Austria	M	2% by 2005; 5.75% by 2010; 10% by 2020
Belgium	T	2.5% by 2005, 5.75% by 2010
Czech republic	T	3.7% by 2005, 5.75% by 2010
Estonia	T	2% by 2005, 5.75% by 2010
Finland	M	2% by 2008, 4% by 2009, 5.75% by 2010
France	M	2% by 2005; 5.75% by 2010; 10% by 2020
Greece	T	0.7% by 2005, 5.75% by 2010
Hungary	T	0.6% by 2005, 5.75% by 2010
Ireland	T	0.06% by 2005 Ireland provides tax exemption within a quota
Italy	T	1% by 2005, 2.5% by 2010
Netherlands	M	2% by 2007, gradually rising to 5.75% by 2010
Latvia	T	2% by 2005, 5.75% by 2010
Lithuania	T	2% by 2005, 5.75% by 2010
Poland	T	0.5% by 2005, 5.75% by 2010
Portugal	T	2% by 2005, 5.75% by 2010
Slovakia	M	2% by 2006, 5.75% by 2010
Slovenia	M	1.2% by 2006, gradually rising to 5% by 2010
Spain	M	3.4% by 2009, rising to 5.83% by 2010
Sweden	T	3% by 2005, 5.75% by 2010
United Kingdom	M	2.5% by 2008, 3.75% by 2009, 5% by 2010
USA	M	2.78% by volume of gasoline consumption in 2006 (4 billion gallons, or 15 GL); 7.5 billion gallons (28 GL) by 2012

Source: GSI (2007). For EU countries with targets not already declared we have assigned an average value corresponding to the EU general target (i.e., Austria and Sweden).

In Brazil support for ethanol production is provided mainly by market regulations, which impose a blending ratio for ethanol with gasoline of 20–25 per cent. There are other forms of incentives such as credit provision for ethanol storage and tax incentives for flex–fuel vehicles whose sales have increased dramatically since their introduction. In Brazil dual plants predominate in which production can be shifted easily from sugar to ethanol production, according to market conditions. Other policies specific to the biofuel sector provide support for distribution and use, support for capital investment and government action to support development and innovation. These forms of support are not considered in our model due to the difficulties this would introduce in terms of consistent data. Agricultural support through payments coupled to production or to land, or providing border protection through high tariffs for agricultural products, may be involved in the total account of support to feedstock production; however, they are also not considered here because they are not specific to final use of crops.

Environmental regulation and technological competitiveness

Since the 1980s, many OECD countries have introduced alternative policy measures aimed at reducing the environmental impacts of economic activity. However, the effects of these policy actions on the economic system are difficult to predict, especially in terms of their effects on the pattern of technological progress.

Many empirical studies analyse the effects of environmental polices on innovation and competitiveness, using different hypotheses and empirical models. Although not exhaustive, two major strands of literature have emerged based on hypotheses that provide some useful insights: the pollution–haven hypothesis, and the Porter and van der Linde hypothesis. These hypotheses are oriented towards investigation of the effects of environmental regulation on international competitiveness and, indirectly, on possible induced technical change. The pollution–haven hypothesis states that the application of more lenient environmental regulation results in reduced production costs for manufactured goods, which improves countries' abilities to export, but also increases domestic pollution emissions (Bommer 1999; Antweiler et al. 2001; Copeland and Taylor 2004; Ederington et al. 2004). The Porter and van der Linde (1995) hypothesis adopts a quite different interpretation of the effects of environmental regulation on dynamic competitiveness theory, deriving from technological innovation linked to stringent environmental standards: the compliance costs related to the introduction of severe environmental regulations should stimulate a country to increase flows of green innovation and become a net exporter of environmental technologies.

Empirical studies investigating the pollution–haven hypothesis do not find fully robust support for this argument (Ederington and Minier 2003; Harris et al. 2002; Jug and Mirza 2005; Levinson and Taylor 2004), while empirical findings for the Porter hypothesis are based mainly on specific industries (Albrecht 1998; Murty and Kumar 2003; Wagner 2003, 2006).

One reason for these unsatisfactory results is the poor indicators for both regulation and environmental innovation (Jaffe et al. 2003, 2005; Jaffe and Palmer 1997; Jaffe et al. 1995). However, the availability of data has increased since the beginning of the 2000s, which has sparked a revival in empirical studies of the existence of specific trade paths related to environmental regulation. In order to produce results valid for the whole economy, some of the more recent contributions have adopted gravity equation models, investigating both the pollution haven (Harris et al. 2002; Jug and Mirza 2005) and the Porter and van der Linde hypotheses (Costantini and Crespi 2008a,b; van Beers and van den Bergh 2003).

We adopt a similar analytical framework, following the original model provided in Costantini and Crespi's (2008a) study, which is based on 20 OECD countries for the period 1996–2006. We adopt this approach for two main reasons. The first is that public support policies for production and consumption of biofuels are very recent – since 2000. A gravity equation enables a wide dataset with a sufficient number of observations to provide statistical robustness to the covariates related to biofuel policies. Secondly, the ultimate aim of this chapter is to formulate some policy advice concerning the capacity of environmental policies (specifically for biofuels) to reinforce international competitiveness claimed by the recent revision to the Lisbon Agenda for the EU, in which sustainability goals are held up as examples of win–win policies, promoting environmental protection and economic development. If the effects of public support policies on biofuels are to divert investments and reduce the competitiveness of energy–saving and renewable energy technologies, this could imply a noticeable conflict between policy actions, especially for the EU.

Empirical model and dataset

The empirical formulation of the gravity equation in this chapter is quite similar in structure to gravity equations used to analyse the impact on trade flows of environmental stringency.

The exporting countries in our analysis (the *i* countries in our gravity equation) are Australia, Austria, Belgium, Canada, Denmark, Finland, France, Germany, Greece, Ireland, Italy, Japan, Netherlands, Norway, Portugal, Spain, Sweden, Switzerland, the UK and the US. The sample for *j* importing countries includes 148 countries (including the OECD countries), and the time period analysed is 1996 to 2006.

The gravity equation analysed in a panel context is formulated as follows:

$$\mathrm{ENEXP}_{ijt} = \alpha + \beta_1 \mathbf{GRAV}_{ijt} + \beta_2 \mathbf{ENV}_{ijt} + \beta_3 \mathbf{BIOF}_{it} + \varepsilon_{ij} \tag{8.1}$$

The dependent variable $\mathrm{ENEXP}_{\mathrm{ijt}}$ is generated by the sum of the bilateral export flows from country *i* to country *j* at time t, of two different aggregations, all expressed in terms of 2000 constant US$ purchasing power parity (PPP) prices: technologies for renewable energies; technologies for energy saving. Export flow data are extracted from the United Nations Conference on Trade and Development (UNCTAD 2008) COMTRADE database, which is based on the Harmonized Commodity Description and Coding System (HS 1996). The typologies of technologies for renewable energies and energy efficiency, excluding technologies for the production of bio–energies, are defined by the OECD (Steenblik 2005a,b), based on the HS 1996 classification. The list proposed by OECD includes all processes and products with the principal purpose of environmental protection, in order to respond to the necessity of an internationally valid definition of environmental goods and services useful in World Trade Organization (WTO) negotiations. The exact definition of these kind of goods is functional to those partners requesting an early specific negotiation agreement, in order to obtain consistent tariff reductions. The sample

Table 8.5 Codes HS 1996, technologies for renewable energies and energy savings

Code	*Description*
Renewable energies	
7321.13	Cooking appliances and plate warmers for solid fuel, iron or steel
7321.83	Non electrical domestic appliances for liquid fuel
8410.11	Of a power not exceeding 1,000kW
8410.12	Of a power exceeding 1,000 kW but not exceeding 10,000 kW
8410.13	Of a power exceeding 10,000 kW. 8410.90 – Parts including regulators
8410.90	Hydraulic turbines and water wheels; parts including regulators
8413.81	Pumps for liquids, whether fitted with a measuring device or not; [Wind turbine pump]
8419.11	Instantaneous gas water heaters
8419.19	Instantaneous or storage water heaters, non–electric — other [solar water heaters]
8502.31	Electric generating sets and rotary converters — Wind powered
8502.40	Electric generating sets and rotary converters [a generating set combining an electric generator and either a hydraulic turbine or a Sterling engine]
8541.40	Photosensitive semiconductor devices, including photovoltaic cells whether assembled in modules or made up into panels; light–emitting diodes
Energy savings and management	
3815.00	Catalysts
7008.00	Multiple–walled insulating units of glass
7019.90	Other glass fibre products
8404.20	Condensers for steam or other vapour power units
8409.99	Parts suitable for use solely or principally with the engines of HS 8407 or 8408; other
8418.69	Heat pumps
8419.50	Heat exchange units
8419.90	Parts for heat exchange equipment
8539.31	Fluorescent lamps, hot cathode
8543.19	Fuel cells
9028.10	Gas supply, production and calibrating metres
9028.20	Liquid supply, production and calibrating metres
9032.10	Thermostats

Source: ENEA (2007) and Steenblik (2005a, 2005b).

in our analysis is restricted to technologies for the energy sector (Table 8.5); thus we merge the classification proposed by Steenblik (2005a,b) with a specific study on environmental technologies provided by the Italian Research Institute for New Technologies, Energy and the Environment (ENEA 2007). This methodological choice derives from the specific purpose of the analysis in this chapter to investigate the role of public support for biofuels on the export dynamics of renewable energies and energy–saving technologies.

The variables we include as independent covariates are aggregated into five groups reported in Table 8.6. This choice is functional to the interpretation of the econometric results, which focus on different aspects of our framework to evaluate the role of the drivers considered here, both separately and together.

The first group (GRAV) collects the variables included in a standard gravity equation model, as income (GDP) and population (POP) for countries *i* and *j* (World Bank, 2009), bilateral geographic distance (GEODIST) between trading partners, following the calculations provided by different studies (CEPII 2006; Mayer and Zignago 2006), total land area as a dimensional variable (LAND), and two dummy variables explaining the existence of past colonial relationships (COL) and geographic contiguity (CONT) which take the value one if the two trading partners are neighbours.[1]

The second group refers to alternative measures of environmental regulation (ENV) represented by a set of indicators, in order to investigate more generally the role of regulation in environmental fields, as a driver of international competitive advantage, and to identify the different impacts related to alternative policy measures. The reason for the choice of these variables is explained in Costantini and Crespi (2008a) and OECD (2008); they represent the environmental protection expenditure attributable to the public and private sectors (PACE) as a percentage of GDP and the share of environmental tax revenues on GDP (ENVTAX). A synthetic measure of environmental regulation (ENVREG) is derived as the sum of the three environmental regulation policies previously described.[2]

In order to test our model accounting for the role of the environmental policies adopted in importing countries, we adopt an indirect measure of environmental stringency as the level of CO_2 emission (expressed as kg per unit of GDP at 2000 constant PPP international $). Since the developing countries (the majority of the 148 importing countries) are excluded from any commitment to the Kyoto Protocol, if they take action towards reducing CO_2 emissions per unit of GDP, their development strategies will be oriented towards energy savings and the adoption of renewable energies, which is an indirect indication that they are adopting environmental regulation.

The third group (BIOF) is related specifically to public support for the biofuel sector. As already mentioned, there is a vast range of public policies that are complementary or substitutes. Here, we consider some specific policy measures based on two criteria: that policy actions are implemented by the whole sample of exporting countries, thus reducing possible biases in the estimation results coming these from lack of data; that policy measures are based on an easily identifiable start date. This results in three types of public support policies.

Table 8.6 Definition of variables

*Variable**	*Definition*	*Source*
Dependent variables		
$ENEXP_{ijt}$	Total bilateral export flows in renewable energies and energy saving technologies (at constant 2000$ PPP) (HS definition Table 8.5) of countries *i* and *j*	UNCTAD–Comtrade
Standard gravity (GRAV)		
$GDP_{i,j,t}$	Natural logarithm of GDP (constant 2000 US$) of country *i* and *j*	World Bank WDI
$POP_{i,j,t}$	Natural logarithm of total population of country *i* and *j*	
$LAND^{j}$	Natural logarithm of land area of country *j* (sq. km)	
GEODISTij	Bilateral geographic distances (CEPII calculation, Mayer and Zignago, 2006)	CEPII
COL_{ij}	Existence of colonial relationships between country *i* and *j* (dummy variable)	
CONT*ij*	Geographic contiguity between country *i* and *j* (dummy variable)	
Environmental regulation (ENV)		
$CO_{2i,j,t}$	Natural logarithm of CO_2 emission (kg per 2000 PPP $ of GDP) of country *i* and *j*	World Bank WDI
$PACE_{it}$	Current environmental protection expenditure (public+industry) as % of GDP	OECD
$ENVTAX_{it}$	Revenues from environmental taxes as % of GDP	
$ENVREG_{it}$	Sum of environmental regulation policies $PACE_{it}$+$ENVTAX_{it}$+$ENVRD_{it}$ (%)	
Support policies for biofuels (BIOF)		
$AHSBF_{it}$	Applied MFN tariff ad valorem for biofuels,[1] weighted with import flows (%)	UNCTAD–TRAINS
$AHSET_{it}$	Applied MFN tariff ad valorem for ethanol,[2] weighted with import flows (%)	
$AHSBD_{it}$	Applied MFN tariff ad valorem for biodiesel,[3] weighted with import flows (%)	
$MANDBF_{it}$	Fuel mandate, targets of blending shares of total consumption (%)	GSI
$EXCBF_{it}$	Arithmetic mean of EXCET and EXCBD (US$ per litre of biofuels)	
$EXCET_{it}$	Value of excise tax reductions for Ethanol or ETBE (US$ per litre of pure ethanol equivalent)	
$EXCBD_{it}$	Value of excise tax reductions for Biodiesel or pure plant oil (US$ per litre of pure biodiesel equivalent)	
$TAXBF_{it}$	Arithmetic mean of TAXET and TAXBD (%)	
$TAXET_{it}$	Share of excise tax reduction for ethanol on total excise tax on gasoline (%)	
$TAXBD_{it}$	Share of excise tax reduction for biodiesel on total excise tax on diesel (%)	
$POLICYBF_{it}$	Arithmetic mean of AHSBF, MANDBF, and TAXBF (%)	

continued ...

Notes to Table 8.6

* Symbols for the identification of countries and time period must be interpreted as follows:

ijt represents the bilateral interaction between exporting and importing countries with a temporal dimension.

ij represents the bilateral interaction between exporting and importing countries without a temporal dimension.

i,j,t represents the value of the variable for country *i* and *j* respectively, with a temporal dimension.

it represents the value of the variable for country *i* with a temporal dimension.

jt represents the value of the variable for country*j* with a temporal dimension.

1 Average weighted tariff for codes (HS 1996): 2207.10 (ethanol), 2905.11 (methanol), 1205.00 (rape or colza seeds, whether or not broken), 1507.10 (crude oil, whether or not degummed), 1511.10 (crude oil), 1512.11 (crude oil), 1514.10 (crude oil).

2 Average weighted tariff for codes (HS 1996): 220710 (ethanol), 290511 (methanol).

3 Average weighted tariffs for codes (HS 1996): 120500 (Rape or colza seeds, whether or not broken), 150710 (crude oil, whether or not degummed), 151110 (crude oil), 151211 (crude oil), 151410 (crude oil).

The first is calculated as tariffs imposed on international import flows of biofuels, split between ethanol and vegetable oils (raw material for biodiesel), derived from the UNCTAD–TRAINS database, expressed in terms of MFN applied duties in ad valorem equivalents. MFN applied tariffs are preferred to bound duties in terms of reducing the biases related to the possibility that bound tariffs for protected sectors are inflated to reap advantages in the WTO negotiations process.[3] There are three variables involved, referring specifically to ethanol and vegetable oils (AHSET and AHSBD, respectively) and to biofuels in general (AHSBF). All tariffs are calculated as averages of the ad valorem equivalent weighted by the corresponding trade flow.

The second is fuel mandates (MANDBF) expressed as a percentage target relative to the corresponding specific fossil fuel (gasoline for ethanol, diesel for biodiesel). In this case we consider only one policy measure for all biofuels (expressed as a simple average of the mandates in the case of two separate targets) because the differences between ethanol and biodiesel are minimal.

The third is linked to excise tax reductions favouring bioethanol and biodiesel consumption. Here, we take the values of the tax reductions (US$ per litre) for ethanol (EXCET), biodiesel (EXCBD), and globally for biofuels (EXCBF). We compute the share of excise tax reduction on total excise tax for the two biofuels separately (TAXET and TAXBD), and generally for biofuels (TAXBF), to homogenize the unit values (percentage) of this specific policy action and others. Data for this policy measure and for fuel mandates are provided by the International Institute for Sustainable Development's Global Subsidies Initiative (GSI).

Lastly, we build a synthetic policy measure (POLICYBF) to assess more generally the impact of public support for biofuels on the competitive advantage of other clean energy technologies. Our variable is based on the arithmetic means of AHSBF, MANDBF and TAXBF, all expressed in percentage terms.

Empirical results

As in previous empirical studies, the gravity equation model provides a good framework of analysis to test the effects of environmental policies in driving technological competitiveness in the energy sector. Table 8.7 reports the results for

Table 8.7 The role of environmental regulation

Dependent variable	*Exports of renewable energies and energy saving technologies (ENEXP)*			
	CO_2	*PACE*	*ENVTAX*	*ENVREG*
GDP_j	0.012	0.056*	0.024	–0.007
	(0.38)	(1.84)	(0.78)	(–0.22)
GDP_i	1.76***	1.47***	2.414***	2.24***
	(30.29)	(25.05)	(40.94)	(37.31)
POP_j	–0.103**	–0.11***	–0.099**	–0.100**
	(–2.35)	(–2.50)	(–2.31)	(–2.32)
POP_i	–0.724***	–0.51***	–1.19***	–1.14***
	(–12.40)	(–8.77)	(–20.47)	(–19.12)
GEODIST	–1.530***	–1.55***	–1.437***	–1.52***
	(–74.52)	(–79.34)	(–73.46)	(–77.58)
COL	1.34***	1.50***	1.34***	1.42***
	(27.34)	(30.57)	(28.01)	(29.21)
CONT	–0.28***	–0.39***	–0.109	–0.294***
	(–3.61)	(–5.06)	(–1.44)	(–3.82)
$LAND_j$	–0.553***	–0.55***	–0.55***	–0.556***
	(–23.09)	(–23.05)	(–23.71)	(–23.49)
CO_{2j}	–0.11***	–0.135***	–0.127***	–0.110***
	(–2.65)	(–3.29)	(–3.13)	(–2.68)
CO_{2i}	–0.317***			
	(–9.54)			
$PACE_i$		0.523***		
		(22.87)		
$ENVTAX_i$			0.550***	
			(37.13)	
$ENVREG_i$				0.183***
				(25.75)
Adjusted R^2	0.69	0.70	0.71	0.70
Obs	24,569	24,569	24,569	24,569

Notes: Statistics for t–Student in parenthesis. *** $p < 0.01$, ** $p < 0.05$, * $p < 0.1$.

a standard gravity equation model augmented with environmental policy variables.[4] In this analysis the market dimension of the exporting country seems to be a more incisive driver of export dynamics than the dimension of the receiving market. Also, the variables related to the proximity of trading countries play a significant role.

The results show that environmental regulation has an impact on renewable energy and energy–saving technologies export flows. The coefficients associated to all the proxies for environmental stringency are statistically significant, and with the expected (positive) sign. For the sample including all countries, this confirms the results obtained in previous studies that consider measures of environmental regulation for European countries only (Costantini and Crespi 2008b).

Note that PACE and ENVTAX – the variables representing market–oriented policy measures – show higher coefficient than the global environmental regulation measure (ENVREG). This confirms the increasing need for an investigation of the real effects of alternative environmental regulation policies on the market, which could constitute an interesting future research task.

After assessing the impact of environmental regulation generally, we introduce a number of variables into the analysis, for policy measures favouring the production and consumption of biofuels. The first three columns in Table 8.8 present the results of the models testing the impact on competition of biofuel tariffs for energy technologies exports. The coefficients are negative and significant, suggesting a relevant influence of such policy measures on the direction of technological change in the energy sector.

This result holds when the share of excise tax reduction for biofuels is considered (columns 4 to 6), since the values of the variables increase with the incentives for the production and consumption of biofuels.[5] Finally we constructed and tested the influence of a synthetic variable related to policies for biofuels (POLICYBF) which considers tariffs, fuel mandate and excise tax reduction: this is significant and has the expected (negative) sign.

Conclusions

The analysis in this chapter tested an empirical model based on a gravity equation, in order to provide evidence of the negative impact produced by public policies supporting the biofuel sector, on the export capacity of selected industrialized countries, for renewable energies (excluding biofuels) and energy-saving technologies.

Our gravity equation model applied to a very specific definition of energy technologies, provides a clear indication that pervasive public policies can be detrimental to investment in and competitiveness of new energy technologies. While countries with stricter environmental standards show higher comparative advantage in the export markets for energy technologies, the introduction of specific public support for the biofuel sector is strongly negatively related to competitiveness internationally.

These results confirm our research hypothesis, which introduces some doubt about the win–win effects related to the policies being implemented by many

Table 8.8 The role of policies for biofuels

Dependent variable	*Exports of renewable energies and energy saving technologies (ENEXP)*						
Biofuels policy	*AHSBF*	*AHSET*	*AHSBD*	*TAXBF*	*TAXET*	*TAXBD*	*POLICYBF*
GDP_j	0.049*	0.053*	0.051*	0.113***	0.083***	0.075**	0.074**
	(1.59)	(1.73)	(1.65)	(3.65)	(2.73)	(2.42)	(2.43)
GDP_i	1.515***	1.47***	1.641***	1.92***	1.94***	1.565***	1.82***
	(25.58)	(24.86)	(26.95)	(29.70)	(30.87)	(25.15)	(28.92)
POP_j	–0.109***	–0.109***	–0.108***	–0.112***	–0.109***	–0.110***	–0.111***
	(–2.51)	(–2.50)	(–2.50)	(–2.59)	(–2.54)	(–2.54)	(–2.57)
POP_i	–0.562***	–0.507***	–0.703***	–0.920***	–0.900***	–0.604***	–0.851***
	(–9.49)	(–8.62)	(–11.46)	(–14.48)	(–14.67)	(–9.74)	(–13.59)
GEODIST	–1.54***	–1.55***	–1.55***	–1.53***	–1.50***	–1.56***	–1.51***
	(–78.04)	(–76.64)	(–79.53)	(–78.85)	(–76.71)	(–79.49)	(–76.98)
COL	1.49***	1.49***	1.49***	1.48***	1.47***	1.49***	1.47***
	(30.46)	(30.45)	(30.40)	(30.30)	(30.27)	(30.52)	(30.19)
CONT	–0.371***	–0.388***	–0.381***	–0.364***	–0.325***	–0.394***	–0.334***
	(–4.81)	(–5.02)	(–4.95)	(–4.75)	(–4.25)	(–5.12)	(–4.35)
$LAND_j$	–0.55***	–0.55***	–0.548***	–0.54***	–0.547***	–0.545***	–0.545***
	(–23.08)	(–23.08)	(–23.11)	(–22.98)	(–23.22)	(–22.97)	(–23.07)
CO_{2j}	–0.130***	–0.135***	–0.134***	–0.180***	–0.163***	–0.149***	–0.152***
	(–3.17)	(–3.28)	(–3.27)	(–4.40)	(–4.00)	(–3.61)	(–3.72)
$PACE_i$	0.518***	0.523***	0.538***	0.490***	0.517***	0.510***	0.494***
	(22.67)	(22.87)	(23.55)	(21.51)	(22.78)	(22.14)	(21.65)
$AHSBF_i$	–0.002***						
	(–6.00)						
$AHSET_i$		–0.001*					
		(–1.47)					
$AHSBD_i$			–0.007***				
			(–10.41)				
$TAXBF_i$				–0.008***			
				(–16.22)			
$TAXET_i$					–0.007***		
					(–20.12)		
$TAXBD_i$						–0.002***	
						(–4.44)	
$POLICYBF_i$							–0.005***
							(–15.05)
Adj. R^2	0.70	0.70	0.70	0.70	0.70	0.70	0.70
Obs	24,569	24,395	24,569	24,569	24,569	24,569	24,569

Note: Statistics for t–Student in parenthesis. *** $p < 0.01$, ** $p < 0.05$, * $p < 0.1$.

industrialized economies, and particularly the EU, to support the diffusion of biofuels. The econometric estimates indicate clearly that technological specialization and export competitiveness in energy technologies can be affected by heavy subsidization of biofuel production and consumption.

The policy implications of this analysis are that the pervasive use of public policies which may be difficult to remove in the future should be avoided: such policies may work to distort energy markets rather than to achieve competitiveness and security of energy supply.

Acknowledgements

The authors gratefully acknowledge the support provided by the research network Enea–Inea–Uniroma Tre on 'Integrating bottom–up and top–down energy models: the case of Gtap–E and Markal–Italy' on which this work is based. Financial support from the Italian Ministry of Education, University and Research (Scientific Research Programme of National Relevance 2007 on European Union policies, Economic and Trade Integration Processes and WTO Negotiations) is also gratefully acknowledged. We are indebted to Jans Horbach, Alessandro Olper, Klaus Rennings, and Markus Wagner for helpful comments and suggestions. The usual disclaimers apply.

Notes

1 In this chapter, we adopted simple distances as our distance measure, in which only one city is necessary to calculate international distances. Simple distances are calculated following the great circle formula, which uses latitudes and longitudes for the most important city (in terms of population) or the official capital of a country (Mayer and Zignago 2006).

2 The two indices (PACE and ENVTAX) represent a sort of market–based instruments quantification in monetary terms.

3 The so–called phenomenon of the water in tariffs corresponds to a wide range between bound duties (those declared to WTO) and applied duties (imposed on importing countries in international trade). For further details, see Bouët et al. (2008).

4 Only fixed effects estimates are reported in Tables 8.7–8.8. The significance of the statistics associated with the Hausman test provides indications that country individual effects are relevant in this kind of analysis, thus, fixed effect estimates are the technique that should be applied. Additional econometric technicalities are used in the estimations, but for the sake of simplicity are not reported in the text: in order to make the model consistent with the standard gravity literature, for instance, we added a full set of year dummies (1996–2006) that proved to be jointly significant, in order to capture the effects related to temporal shocks. We also included dummies specific to each *i*–th or *j*–th country, and dummies representing the interaction between countries and time periods, as suggested by Baldwin and Taglioni (2006), but the results are barely affected by these alternative specifications. We checked for autocorrelation by including the lagged value of the dependent variable, but the coefficients are not statistically robust. We included several regional dummies, but the only one with statistically robust coefficients is related to the importing country being a member of the OECD. Finally, we tested different temporal lags associated with the environmental and biofuels policy variables and found the results were still robust

with zero, one and two temporal lags. We report the estimates with zero lags in order to not lose observations.

5 Table 8.8 reports the results for several models, which include all the alternative variables explaining support for both bioethanol and biodiesel, and in all cases the coefficients are negative and statistically robust.

References

Albrecht, J. (1998) 'Environmental Costs and Competitiveness. A Product–Specific Test of the Porter Hypothesis', Working Paper 50, University of Ghent, Ghent.

Antweiler, W., Copeland, B.R. and Taylor, M.S. (2001) 'Is free trade good for the environment?', American Economic Review, 91: 877–908.

Baldwin, R. and Taglioni, D. (2006) 'Gravity for Dummies and Dummies for Gravity Equations'. National Bureau of Economic Research, NBER Working Paper 12516, Cambridge, MA: NBER.

Bommer, R. (1999) 'Environmental policy and industrial competitiveness: The pollution–haven hypothesis reconsidered', Review of International Economics, 7: 342–55.

Bouët, A., Decreux, Y., Fontagné, L., Jean, S. and Laborde, D. (2008) 'Assessing applied protection across the world', Review of International Economics, 5: 850–63.

CEPII (Centre d'Etudes Prospectives et d'Informations Internationales) (2006), Dataset of Distances Measures, Paris: CEPII.

Copeland, B.R. and Taylor, M.S. (2004), 'Trade, growth, and the environment', Journal of Economic Literature, 42: 7–71.

Costantini, V. and Crespi, F. (2008a) 'Environmental regulation and the export dynamics of energy technologies', Ecological Economics, 66: 447–60.

Costantini, V. and Crespi, F. (2008b) 'Environmental institutions and the trade of energy technologies in Europe', International Journal of Global Environmental Issues, 8: 445–60.

Ederington, J. and Minier, J. (2003) 'Is environmental policy a secondary trade barrier? An empirical analysis', Canadian Journal of Economics, 36: 137–54.

Ederington, J., Levinson, A. and Minier, J. (2005) 'Footloose and pollution–free', Review of Economics and Statistics, 87: 283–306.

ENEA (2007), Energy and Environment Report 2006, Rome: ENEA (in Italian).

EU (2003), 'Directive 2003/30/EC of the European Parliament and of the Council of 8 May 2003 on the promotion of the use of biofuels or other renewable fuels for transport', Official Journal of the European Union L 123/42, Brussels: EU.

EUROSTAT (2007) Energy, Transport and Environment Indicators, Eurostat Pocketbooks, Luxemburg: Eurostat.

GSI (2007) Biofuels – At What Cost? Government Support for Ethanol and Biodiesel in the Selected OECD Countries, Global Subsidies Initiative, Winnipeg, Manitoba, Canada: International Institute for Sustainable Development.

Harris, M.N., Kónya, L. and Mátyás, L. (2002) 'Modelling the impact of environmental regulations on bilateral trade flows: OECD, 1990–1996', World Economy, 25: 387–405.

IEA (2007) World Energy Outlook 2007, Paris: IEA.

Jaffe, A.B. and Palmer, K. (1997) 'Environmental regulation and innovation: A panel data study', Review of Economics and Statistics, 79: 610–19.

Jaffe, A.B., Peterson, S.R., Portney, P.R. and Stavins, R.N. (1995) 'Environmental regulation and the competitiveness of US manufacturing: What does the evidence tell us?', Journal of Economic Literature, 33: 132–63.

Jaffe, A.B., Newell, R.G. and Stavins, R.N. (2003) 'Technological change and the environment', in K.G. Mäler and J.R. Vincent (eds), Handbook of Environmental Economics, vol. 1, Amsterdam: Elsevier, 461–516.

Jaffe, A.B., Newell, R.G. and Stavins, R.N. (2005) 'A tale of two market failures: Technology and environmental policy', Ecological Economics, 54: 164–74.

Jug, J. and Mirza, D. (2005) 'Environmental regulations in gravity equations: Evidence from Europe', World Economy, 28: 1591–615.

Levinson, A. and Taylor, S. (2004) 'Unmasking the Pollution Haven Effect', NBER Working Paper 10629, Cambridge, MA: NBER.

Mayer, T. and Zignago, S. (2006) Notes on CEPII's Distances Measures, Paris: Centre d'Etudes Prospectives et d'Informations Internationales (CEPII).

Murty, M.N. and Kumar, S. (2003) 'Win–win opportunities and environmental regulation: Testing of porter hypothesis for Indian manufacturing industries', Journal of Environmental Management, 67: 139–44.

OECD (2005) Agricultural Market Impacts of the Future Growth in the Production of Biofuels, OECD, Doc AGR/CA/APM(2005)24/Final, Paris: OECD.

OECD (2008) Environmental Policy, Technological Innovation and Patents, Paris: OECD.

Porter, M.E. and van der Linde, C. (1995), 'Toward a new conception of the environment–competitiveness relationship', Journal of Economic Perspectives, 9: 97–118.

Steenblik, R. (2005a) 'Environmental Goods: A Comparison of the APEC and the OECD Lists', OECD Trade and Environment Working Paper 04, Paris: OECD.

Steenblik, R. (2005b) 'Liberalisation of Trade in Renewable–energy Products and Associated Goods: Charcoal, Solar Photovoltaic Systems, and Wind Pumps and Turbines', OECD Trade and Environment Working Paper 07, Paris: OECD.

Tokgoz, S., Elobeid, A., Fabiosa, J., Hayes, D.J., Babcock, B., Tun–Hsiang, A., Yu, E., Dong, F., Hart, C.E. and Beghin, J.C. (2007) 'Long–Term and Global Trade–offs between Bio–Energy, Feed, and Food', Center for Agricultural and Rural Development, Iowa State University, Selected Paper for American Agricultural Economics Association Annual Meeting, Portland, OR, 29 July–1 August.

UNCTAD (United Nations Conference on Trade and Development) (2008) COMTRADE, On–line Database, New York: UN.

Van Beers, C. and van den Bergh, J.C.J.M. (2003) 'Environmental regulation impacts on international trade: Aggregate and sectoral analyses with a bilateral trade flow model', International Journal of Global Environmental Issues, 3: 14–29.

Wagner, M. (2003), 'The Porter Hypothesis Revisited: A Literature Review of Theoretical Models and Empirical Tests', Working Paper, Centre for Sustainability Management, Universitat Luneburg.

Wagner, M. (2006) 'A comparative analysis of theoretical reasoning and empirical studies on the Porter hypothesis and the role of innovation', Zeitschrift für Umweltrecht und Umweltpolitik, 3: 349–68.

World Bank (2009) World Development Indicators (WDI), On–line Database, Washington DC: World Bank.

9 Environmental impacts of personal mobility

Exploring an Austrian EKC

Michael Getzner

Introduction

Traffic (the transportation of people, goods and raw materials) contributes significantly to the emission of air pollutants as well as GHG (greenhouse gases), such as carbon dioxide (CO_2). For instance, the transport sector in Austria currently emits GHG adding up to nearly 23 Mtons of CO_2 equivalents. Transportation is the highest emitting sector in the Austrian economy with a share of 36 per cent of all GHG emissions. The large increase in GHG emissions since 1990 is mainly due to increased fossil fuel use by passenger and freight road transport (Umweltbundesamt 2008). Furthermore, the transport infrastructure, particularly roads and railways, and to a lesser extent transport by ship, may pose threats to the functioning of ecosystems by splitting coherent ecosystems into smaller areas with potentially important negative impacts on the viability of species (Löfvenhaft *et al.* 2004). Other environmental impacts of traffic include noise and other pollutants such as NO_x (nitrogen oxide) and particulates (Umweltbundesamt 2008).

Austria's CO_2 emissions balance shows that one of the main contributors to the country's failure to achieve the Kyoto targets (for Austria: a reduction of 13 per cent based on 1990 emissions) is road transport, and especially private cars and large trucks.

For Austria, the EKC (environmental Kuznets curve) hypothesis has already been tested, for example, by Friedl and Getzner (2003), for total CO_2 emissions in the Austrian economy. The most important driving force of CO_2 emissions is shown to be economic growth; there was no inverted U-shape to the EKC. Rather, CO_2 emissions can be described best by a cubic function corresponding to an N-shaped curve, with the oil-price shock of the mid-1970s dampening growth in CO_2 emissions. However, since the late 1990s, CO_2 emissions have begun to grow significantly, and the latest projections indicate that Austria will not fulfill the Kyoto target: current emissions are about 20 per cent higher than in 1990 (rather than 13 per cent lower). The focus in this chapter is on a detailed examination of the drivers of total growth in passenger transport (measured in passenger-kilometres), and growth of private cars, and CO_2 emissions from private car use. Our analysis uses a transport indicators database with a long time span, that is,

1950 to 2005. Data are available for this period on private transport (passenger-kilometres and CO_2 emissions) for all transport options (road, private and public, railways, air transport) as well as commercial transport (in tons and kilometres, for road, railway and air transport). In order to focus on the decisions made by private households, we look at passenger transport and ignore the determinants of freight transport (Getzner 2008).

Determinants of personal mobility demand: recent studies

Income (personal income, per capita GDP) is generally considered to be the main determinant of personal mobility, and specifically ownership and use of private cars. Income is often identified as the main determinant of car ownership (Dargay and Gately 1997). However, the effect of income on car ownership is not linear. For low earners, cars can be considered a luxury good with an income elasticity above unity; as income increases, elasticity decreases to unity or even less. Also, car ownership responds more positively to rising income than to falling income. Satiation in terms of the number of cars per capita therefore seems to play an important role in explaining mobility demand (Dargay 2001). Other factors include the price of cars and gasoline (costs of travel), road infrastructure and population density (Clark 2007).

Actual car use is also determined by the above factors – income, prices, infrastructure and population. For instance, Wohlgemuth (1997) reviews some 150 papers on transport energy demand. While these papers cover a wide variety of modelling approaches and results, income and price are the most important determinants of energy demand for transportation (Dargay 2007).

These empirical results indicate that the EKC for personal mobility and its related environmental impacts cannot be estimated in an unadjusted model that takes account only of income as the main underlying determinant. As Auci and Becchetti (2006) suggest, estimations of an EKC for CO_2 emissions should also account for energy prices, environmental policies and the rent-seeking behaviour of interest groups, technological change and international trade. In the transportation sector, energy prices and technological change are of particular importance. An inverted U-shaped EKC is generally hypothesized for a number of pollutants (Copeland and Taylor 2004; Dasputa *et al.* 2002; Dinda 2004, 2005), although some have criticized the EKC relationship between income and emissions (Stern 2004). Bartz and Kelly (2008) test this theoretical concept and find only weak evidence of an inverted U-shape. On the other hand, they find evidence of a structural break in preferences in the early 1970s, an interesting empirical result congruent with many other studies (Friedl and Getzner 2003). Bartz and Kelly (2008) provide a survey of the literature giving theoretical explanations for an inverted U-shape EKC. With respect to passenger transport, the most important seem to be: abatement costs, institutional strength and environmental preferences. The convexity of abatement costs, and the fact that environmental quality is a normal good mean that pollution reduces after a certain threshold income level beyond which it is worthwhile to invest in abatement activities. The abatement costs for CO_2 in cars, in fact, may be convex since the cheapest options, e.g. switching fuels, have largely been exploited and more costly options (e.g. electronically managed combustion processes; diesel engine catalytic converters) are required. In terms of institutional strength, increasing incomes result in stronger legal systems and institutions, which leads to the implementation and execution of stricter rules, such as technological standards or other regulation (e.g. speed limits), and in relation to environmental preferences, if environmental quality becomes a sufficiently strong preference, and the marginal utility of consumption (income) falls, clean technologies will become optimal.

Regarding demand for gasoline for by private cars, Storchmann (2005) introduces some interesting arguments. While income and price are generally accepted to be crucial determinants of gasoline demand, railway prices, population density and road infrastructure can also be decisive. Storchmann goes on to explore the importance of income distribution, and suggests that in cross-country comparisons, countries with greater inequalities in income distribution have smaller car fleets and lower gasoline demand.

In addition, technological progress leads to a reduction in energy intensity (energy use) while the rebound effect leads to higher energy demand (e.g. demand for larger and heavier cars) and questions whether policies that concentrate solely on increasing energy efficiency will be sufficient to achieve a sustainable development path (see Hanley *et al.* 2008 and Kwon 2005, who employ an IPAT model to identify the main components of CO_2 emissions). This argument is discussed by Ruzzenenti and Basosi (2008), who argue that technological progress is more often used to increase engine power than to improve energy efficiency.

Empirical analysis

Some descriptive evidence

For passenger transport, especially private cars, the time series of passenger-kilometres exhibits a clear upward trend, from around 19 bn passenger-kilometres in 1950 or some 2,700 kilometres per capita per year, to around 103 bn passenger-kilometres in 2006 or some 12,400 kilometres per capita per year. Figure 9.1 shows the development over time; transport in private cars has contributed the most to the increase in distances travelled, while public transport (trains, buses, trams, subways) has increased to a much smaller extent.[1] Together with the increase in private passenger mobility, the environmental impacts of private transportation have also increased dramatically. While it is not straightforward to measure the impact of road traffic on biodiversity (e.g. fragmentation of ecosystems or destruction of wetlands) the negative environmental impacts of air pollution are well known.

Figure 9.2 depicts one of the main environmental impacts of private car use in Austria. Since 1990, GHG emissions from private car use have increased roughly proportionally to GDP.[2] It is difficult to consider a longer time span because emissions were not recorded coherently, using a common methodology, and were not attributed to the different transportation modes (road, rail, water, etc.) or

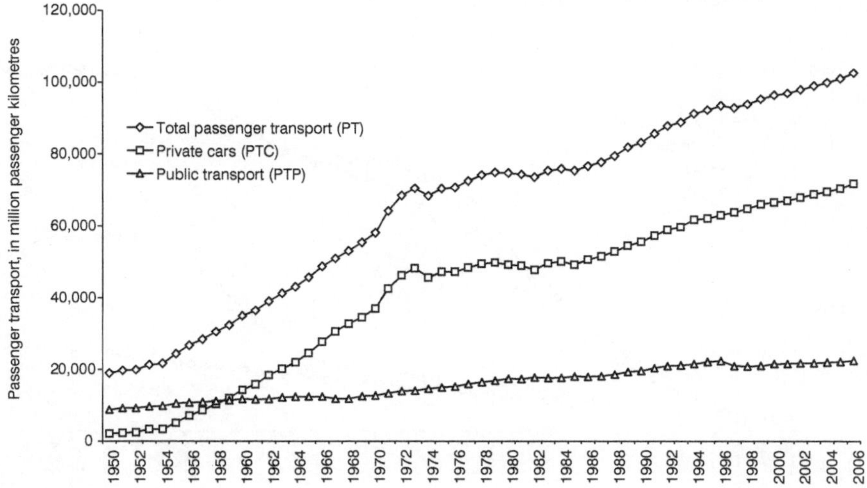

Figure 9.1 Passenger transport in Austria 1950–2006 (in m passenger-kilometres)
Source: Statistik Austria / Umweltbundesamt; for a detailed description of the variables PT, PTC, PTP see Table 9.1.

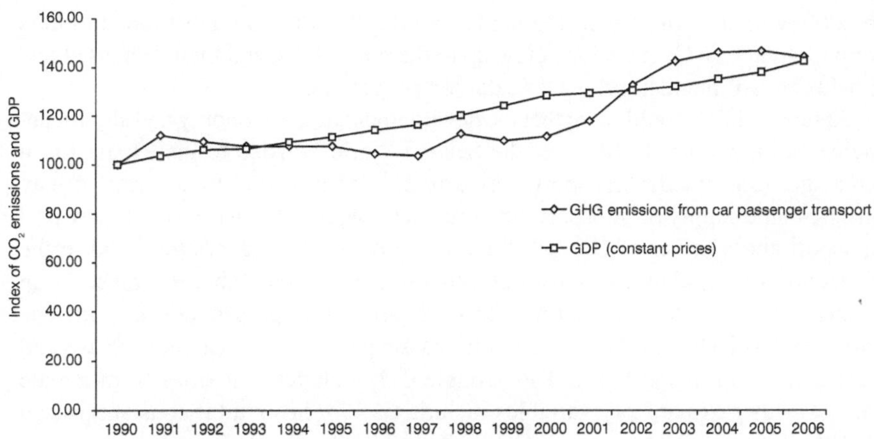

Figure 9.2 GHG emissions (CO_2 equivalents) from passenger transport in Austria and GDP (at constant prices) 1990–2006 (index 1990 = 100)
Source: Statistik Austria / Umweltbundesamt, WIFO; author's calculations.

motives (private or commercial). However, we have time series for CO_2 emissions, particulates and NO_x from 1950. If we look at total CO_2 emissions from private and commercial passenger road transport, we can see that there is a dramatic increase. Figure 9.3 shows that CO_2 emissions increased at a much higher rate than GDP. Particulates grew proportionally to GDP until the late 1980s. Beginning with the boom in the use of diesel engines (due to lower fuel tax in Austria for diesel compared with gasoline), particulate emissions increased dramatically up

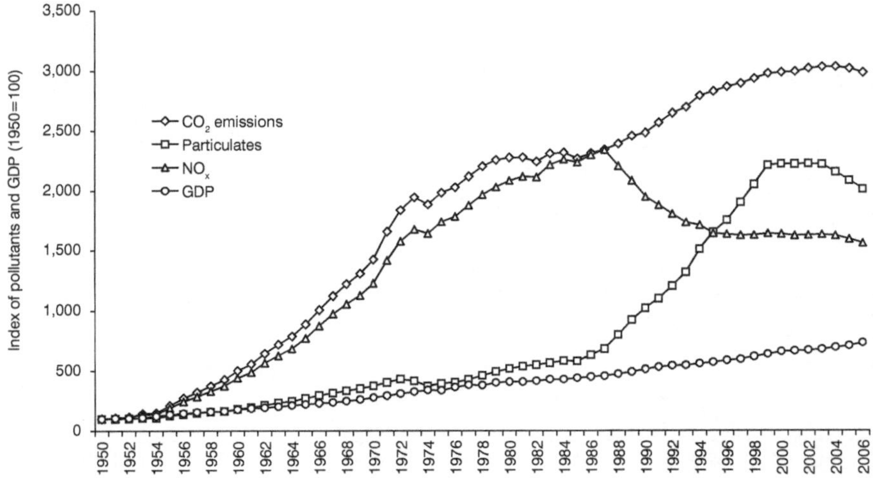

Figure 9.3 Emissions (CO_2, air pollutants) from passenger road transport in Austria and GDP (at constant prices) 1950–2006 (index 1990 = 100)
Source: Statistik Austria / Umweltbundesamt, WIFO; author's calculations.

to 2000, with a slight decreasing tendency after that due to innovations in filters for particulates. NO_x emissions grew up to the mid-1980s, and then decreased due to widespread (mandatory) use of catalytic converters.

At first sight, it would seem that the development of the economy and significant increases in income (GDP) after the Second World War led to the expansion in passenger (and goods) transport. The driving forces behind transportation may also lie in the supply of transport infrastructure, which is both a cause of increased transport and a driver of GDP, in the relative price of transportation, and in the dispersion of populations as well as economic activities. Table 9.1 presents an overview of the potential determinants of the growth of passenger transport. The variables PT, PTC and PTP denote total passenger transport, private car use and public transport, respectively. The variable CO_2 includes emissions from private car use. Other explanatory variables include GDP per capita (at constant prices) (YRPOP), deviations from trend-GDP (YGAP), fuel prices (as a rough measure of the costs of car use) (FUELPR), the length of the federal road network (ROADS), and the concentration of the Austrian population (measured as a Herfindahl index of concentration) (HERF).

Econometric approaches

The aim of this chapter is to reject or confirm the existence of an EKC for passenger transport (total, and private cars only) and CO_2 emissions (from private car use) for Austria. We have complete datasets for the period 1959 to 2006. The methodology used is described below.

First, we describe the characteristics of the time series of dependent and explanatory variables in terms of stationarity and order of integration. It seems

Table 9.1 Variables for estimating an EKC for passenger transport and CO2 emissions from private car use

Dependent variables	*Description*
PT	Total passenger transport (cars, public transport), in m passenger-kilometers
PTC	Passenger transport (private cars), in m passenger-kilometers
PTP	Passenger transport (public transport), in m passenger-kilometers
CO_2	CO_2 emissions from private car use, in 1,000 tons
Explanatory variables	
YRPOP	Per-capita GDP (constant 2000 prices)
YGAP	Deviation from trend-GDP (in per cent)
FUELPR	Gasoline price (constant 2000 prices)
ROADS	Length of the road network
HERF	Concentration index of the Austrian population (0=evenly distributed over space; 1=concentrated in one spot)

that most of the variables are I(1) variables (non-stationary in level, but stationary in first differences). Simply regressing the dependent variables on the set of explanatory variables may lead to spurious regression results. Therefore, second, we test for cointegration between these variables. We test for bilateral cointegration by applying the Durbin–Watson test of cointegration and by exploring the stationarity of the residuals of the cointegrating equations. In addition, we employ the Johanson–Juselius procedure to test for multilateral cointegration between variables. Third, based on these results, we present the results of different estimation approaches before employing a ML (maximum likelihood) estimation procedure to account for potential heteroscedasticity of the variables. Testing both ordinary least squares (OLS) and two-stage least squares (2SLS) estimations for homoschedasticity indicates that the variance of the variables differs over time across the observation period. We therefore estimate our models by employing the GARCH (general autoregressive conditional heteroscedasticity) approach, and accounting for Heteroscedasticity-Consistent Standard Errors (HCSE).

It has been suggested that the linear models explaining the relation between income and environmental pressure cannot completely capture a potentially inverted U-shaped EKC. We therefore also test our models for the appropriate functional specification. We begin with the simple linear model assuming no underlying EKC. Then we test for the quadratic functional form assuming the traditional EKC functional form. Finally, in line with some of the recent EKC studies (Friedl and Getzner 2003) that show that a cubic form (N-shaped EKC) would represent a more realistic picture of the income–emission relationship; we conduct the appropriate test.

Stationarity, cointegration and the causality of transport demand and GDP

Employing the ADF (Augmented Dickey–Fuller) and the PP (Phillips–Perron) test for stationarity, the estimation results[3] suggest that the variables (Table 9.1), both dependent and explanatory, are mostly non-stationary in level, but stationary in their first difference (I(1)). Only the time series of fuel prices (augmented by purchasing power) is clearly stationary.[4] However, the picture is not that clear cut, since only the length of the road network exhibits a clear and significant upward trend, while the other variables show different trends and even drifts.

Based on these results, we test the non-stationary variables for cointegration in a two-step procedure. First, the bilateral (pair-wise) cointegration equation is tested for stationarity of residuals, and the Durbin–Watson statistic is compared with the critical values indicating cointegration. Without going into detail,[5] a clear bilateral cointegrating relation can only be detected between passenger transport on the one hand, and CO_2 emissions and ROADS on the other. This result is interesting and important since transport policies have always been justified by the claim that roads (or the provision of infrastructure) react only to demand in terms of an increase in passenger transport. Together with the results of the causality tests (see below), we can detect a parallel movement between transport demand and road provision.

Second, the Johansen–Juselius (1990) cointegration tests are applied to explore potential multilateral cointegration relations. The multilateral cointegration tests indicate more cointegrating relations than suggested by the bilateral tests discussed above. Depending on the assumption of the VAR system, up to three cointegrating relations are suggested by the results, indicating a more complex structure of relations between variables, probably between passenger transport (emissions) and income, road infrastructure, and the concentration of population. This is often considered to be among the most important results in terms of a meaningful EKC concept (Galeotti *et al.* 2006).

In order to corroborate cointegration and the potential causality of the variables, a Granger causality test was applied and rejected the hypothesis (H_0) of no causality between the variables YRPOP and ROADS, and between PT and ROADS. Some indication of causality – though not significant at the 5 per cent level of significance – was revealed between YRPOP on the one hand, and HERF and PT, and also between HERF and ROADS. These relations may be upheld on the following grounds. The driving force behind passenger transport (PT) might be per-capita GDP (YRPOP), while the supply of transport infrastructure (ROADS) and the increasing dispersion of the population (HERF) might also be important. Both variables are influenced by increasing income (GDP). However, increasing passenger transport also demands more infrastructure, and vice versa. This point was underlined by Tapio (2005), who refers to infrastructure policy as a self-fulfilling prophecy. Based on economic growth forecasts, road transport is forecast to increase, leading to congestion on the existing road network; hence, more roads/lanes must be provided, which means road transport increases, leading to an even greater need for extensions to the road network.

Empirical approaches to estimating an EKC

The results of our exploration of the determinants of total passenger transport are presented in Table 9.2 (Est. 1 and 2). Est. 1 presents the results for the linear functional form (upper half of Table 9.2), while Est. 2 includes the results for the assumption of a cubic form. Est. 1 shows that the (long-run) driving force behind total passenger transport is personal income (YRPOP), while short-term fluctuations also lead to short-term changes in passenger transport (YGAP). The length of the road network (ROADS) increases passenger transport significantly, while the dispersion of the population also contributes positively to increased passenger transport. Fuel prices (FUELPR) are significantly negatively correlated to passenger transport.

As the estimations in the lower part of Table 9.2 suggest, there is significant indication of a cubic functional form of the EKC for passenger transport in Austria. Testing for a quadratic functional form leads to insignificant income terms. Est. 2 is clearly the estimation with the better statistical fit compared with Est. 1. The cubic YRPOP term is highly significant and, except for ROADS, the coefficients of all the other variables have the expected signs and significance. The cubic functional form, taken together with a Chow breakpoint test assuming a structural break in the mid-1970s due to the oil price shock, suggests that significant energy efficiency measures and changes in modal split took place after the oil price shock, while increased income especially in the last decade, (over-)compensated for these gains in energy efficiency. Alternative estimations taking into account a time dummy variable from the mid-1970s onward, does not produce satisfactory results.

The next step consists of taking only one important mode of transportation, private car use, as the variable to be explained. The estimations suggest similar relations between private car use and the explanatory variables as found above. Est. 3 presents the results for the linear functional form, Est. 4 includes the cubic form. Again, the maximum-likelihood approach performs best, with long-term income (per-capita GDP) and short-term deviations from trend-GDP being important for explaining private car use. The length of the road network (ROADS) and the dispersion of the population (HERF) are the significant explanatory variables once again. As discussed above, the cubic form of an EKC may also be reasoned by the estimations. The results suggest an N-shaped form for determination of private car use by per capita income.

Finally, we concentrate on one significant environmental effect of passenger transport, and of private car use in particular. CO_2 emissions in Austria are caused to a significant degree by the use of private cars for passenger transport, especially by commuters. As was the case with the two other dependent variables, the results suggest that the main driving force behind CO_2 emissions is income; Est. 5 shows the linear relationship, Est. 6 indicates the cubic form.

Table 9.2 Estimating an EKC for passenger transport in Austria: total passenger transport

	Estimate 1	
Variable: PT	*Coefficient*	*z-Statistic*
C	298,577.10	1.07
YRPOP	1,058.41	3.81***
YGAP	–26.01	–156.23**
ROADS	2.95	21.84***
HERF	–8,480.56	–1.83*
FUELPR	–4,616.11	–2.45**
AR(1)	0.99	134.43***
Adjusted R-squared	0.99	
S.E. of regression	1,229.32	
Log likelihood	–396.49	
F-statistic	1,360.81***	
Durbin-Watson stat.	1.43	
Estimation	ML-GARCH	
Instruments (2SLS)		
Period (*n*)	1959–2006 (48)	
	Estimate 2	
Variable	*Coefficient*	*z-Statistic*
C	–499,689.1	–94.17***
YRPOP	22,138.29	51.11***
$YRPOP^2$	–825.94	–64.46***
$YRPOP^3$	11.63	86.55***
YGAP	–440.72	–8.83***
ROADS	0.05	0.22
HERF	26,472.31	62.83***
FUELPR	–9,556.76	–4.55***
AR(1)	0.77	16.80***
Adjusted R-squared	0.99	
S.E. of regression	1,190.23	
Log likelihood	–395.20	
F-statistic	1,188.13***	
Durbin-Watson stat.	1.49	
Estimation	ML-GARCH	
Instruments (2SLS)		
Period (*n*)	1959–2006 (48)	

(*continued*)

Table 9.2 (continued)

	Estimate 3	
Variable: PTC	*Coefficient*	*z-Statistic*
C	307,756.2	0.36
YRPOP	918.18	5.08***
YGAP	–17.05	–21.24***
ROADS	2.42	7.17***
HERF	–4,685.39	–2.73***
FUELPR	–6,906.35	–1.97**
AR(1)	0.99	83.70***
Adjusted R-squared	0.99	
S.E. of regression	1,200.34	
Log likelihood	–395.63	
F-statistic	974.86***	
Durbin-Watson stat.	1.42	
Estimation	ML-GARCH	
Instruments (2SLS)		
Period (*n*)	1959–2006 (48)	
	Estimate 4	
Variable	*Coefficient*	*z-Statistic*
C	–420,915.50	–5.60***
YRPOP	22,828.39	79.58***
$YRPOP^2$	–938.25	–7.99***
$YRPOP^3$	14.03	96.98***
YGAP	–375.42	–6.85***
ROADS	0.06	0.10
HERF	20,085.98	4.41***
FUELPR	–10,399.39	–5.53***
AR(1)	0.81	13.76***
Adjusted R-squared	0.99	
S.E. of regression	1,100.13	
Log likelihood	–392.69	
F-statistic	950.37	
Durbin-Watson stat.	1.47	
Estimation	ML-GARCH	
Instruments (2SLS)		
Period (*n*)	1959–2006 (48)	

Table 9.2 (continued)

	Estimate 5	
Variable: CO_2	*Coefficient*	*z-Statistic*
C	71,834.05	2.25**
YRPOP	181.75	4.76***
YGAP	–18.03	–7.86***
ROADS	–0.07	–5.66***
HERF	–4,494.73	–2.01***
FUELPR	–1,134.50	–3.85***
AR(1)	1.39	14.30***
AR(2)	–0.42	–4.65***
Adjusted R-squared	0.99	
S.E. of regression	147.52	
Log likelihood	–293.44	
F-statistic	1,591.97***	
Durbin-Watson stat	1.70	
Estimation	ML-GARCH	
Instruments (2SLS)		
Period (*n*)	1959–2006 (48)	

	Estimate 6	
Variable	*Coefficient*	*z-Statistic*
C	–79,088.04	–2.39**
YRPOP	3,125.52	6.52***
$YRPOP^2$	–102.40	–6.42***
$YRPOP^3$	1.17	4.79***
YGAP	–70.92	–4.44***
ROADS	–0.05	–0.28
HERF	4,037.37	2.06**
FUELPR	–1,176.90	–3.07***
AR(1)	0.83	13.06***
Adjusted R-squared	0.99	
S.E. of regression	176.09	
Log likelihood	–307.32	
F-statistic	1,120.959***	
Durbin-Watson stat.	1.47	
Estimation	ML-GARCH	
Instruments (2SLS)		
Period (*n*)	1959–2006 (48)	

Policy scenarios

The models estimated here can be used to design illustrative policy scenarios. Figures 9.4 to 9.6 show potential development of transportation and CO_2 emissions from 2007 to 2020, and highlight the importance of GDP and fuel prices as the main determinants of future car use as well as CO_2 emissions from car use. All three figures include forecasts for two scenarios, which are based on stable relations between explanatory and dependent variables. The two scenarios are built on the assumption that GDP develops according to an autoregressive trend model. We assume, therefore, that future per-capita income will increase by roughly 1.7 per cent per annum, which is a conservative estimate since average GDP growth has been around 2.3 per cent since the mid-1970s. Furthermore, the two scenarios vary in terms of their assumptions about future fuel prices. While holding population dispersion and the length of the road network constant, scenario I assumes that fuel prices remain constant at the 2008 level of around €1.5 per litre. Scenario II assumes that fuel prices increase annually by €0.1 per litre, reaching €2.8 per litre by 2020. These scenarios are built, therefore, on the empirical result that income and fuel prices are the main determinants of use.

Figures 9.4 to 9.6 show clearly that – based on our estimated models – growth in income as the main driver of private car use will lead to a significant increase in passenger transport. In scenario I, total passenger transport (Figure 9.4) will increase from roughly 102 bn passenger-kilometres to about 126 bn passenger-kilometres in 2020. Such an increase can be considered the 'business-as-usual' scenario because passenger transport grew by about 1.5 per cent over the last few decades. Scenario II, with ever-increasing fuel prices, arrives at 114 bn passenger-kilometres in 2020. The policy option of increasing fuel prices, therefore, suggests that environmental taxation could play an important role as a major environmental policy instrument.

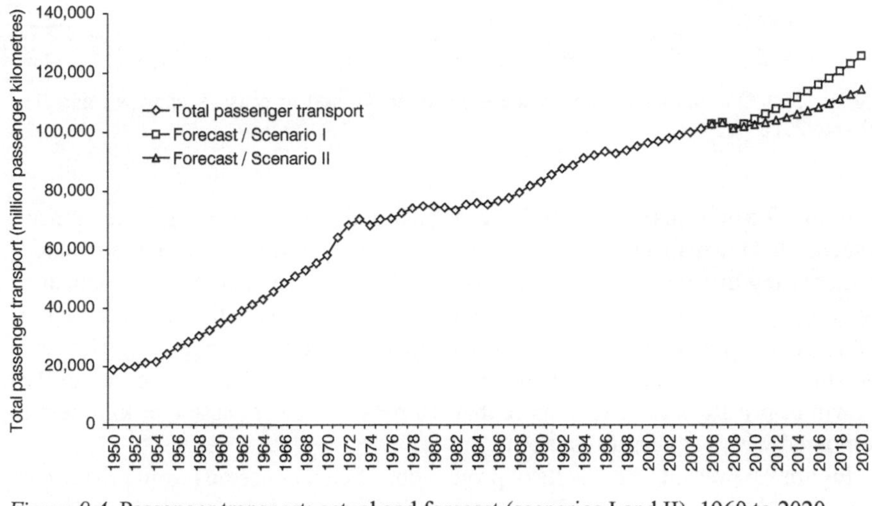

Figure 9.4 Passenger transport: actual and forecast (scenarios I and II), 1960 to 2020

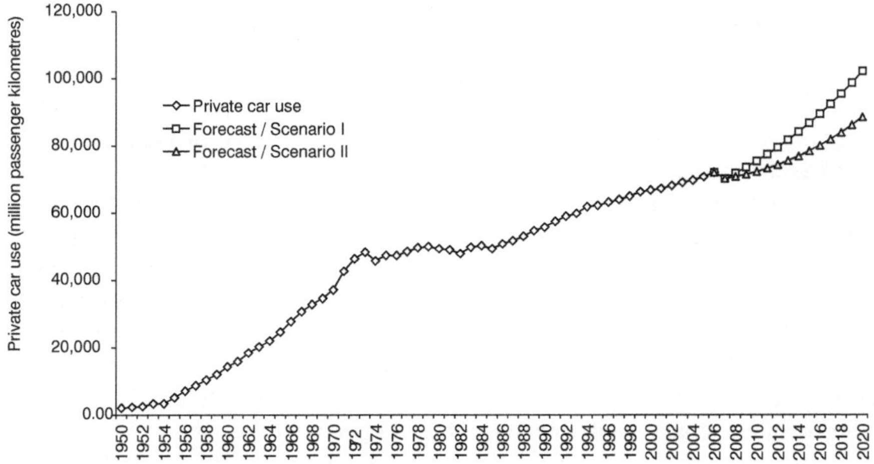

Figure 9.5 Private car use: actual and forecast (scenarios I and II), 1960 to 2020

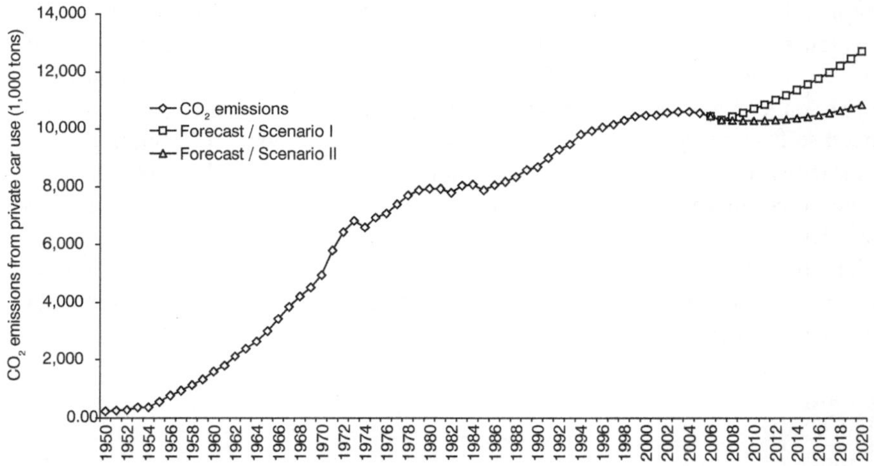

Figure 9.6 CO_2 emissions from private car use: actual and forecast (scenarios I and II), 1960 to 2020

Figure 9.5 presents the results for scenarios I and II in the case of private car use. It seems that the main share of the increase in passenger transport can be attributed to increasing private car use. In scenario I, private car use increases dramatically by roughly 30 per cent, which means that without restrictive environmental policies, other transport modes will lose market share to the 'benefit' of private car use. In scenario II, increasing fuel prices leads to a substantial dampening in the growth of private car use with more than 10 per cent fewer passenger-kilometres compared with business as usual.

Figure 9.6 presents the results of projections of CO_2 emissions from private car use. The relative increase in CO_2 emissions is lower compared with the increase in the volume of passenger transport. In scenario I, CO_2 emissions increase by roughly 20 per cent, while in the restrictive scenario II, the increase amounts to only about 4 per cent. The disproportionate increase in CO_2 is an indication of the improved fuel efficiency of vehicles. Therefore, the environmental impacts of the increase in transport are mediated by technological progress.

Economic growth and fuel prices can be identified as the main driving forces of passenger transportation. However, the scenarios described above are built on a stable relation between economic growth and demand for passenger transport. While no turning point towards reducing passenger transport is detected, the Lucas (1976) critique is relevant since the parameters of the estimated models may change after the implementation of climate change policies. Therefore, the projections within the scenarios are an illustration only that the income–mobility demand relationship, which also forms the basis of any rebound effect (Spielmann *et al.* 2008) in environmental economics modelling, needs to be changed, and that much more than strict environmental taxation is needed to reduce or cap emissions. The scenario results suggest that, at least for emissions, some cap and trade mechanism for reducing CO_2 emissions (Raux and Marlot 2005; Wadud *et al.* 2008) needs to be implemented in order to increase significantly the planning reliability of environmental policies such as meeting internationally binding GHG emission reduction targets.

Conclusions

The development of passenger transport, and especially of private car use, is one of the main contributors to environmental pressure in terms of emissions, noise and loss of biodiversity. In Austria, CO_2 emissions have increased significantly over the last 50 years, reflecting the rising importance of the transport sector. In fact, the main economic activity contributing to higher emissions in 2009 is the use of private cars. It is of great interest, therefore, to explore the determinants of passenger transport – and private car use in particular.

The empirical and theoretical literature on the relation between income (measured e.g. by personal income or per-capita GDP) and environmental impacts suggests an inverted U-shaped relation in the case of an EKC. While there are single pollutants that develop according to the EKC concept (NO_x and SO_2), some of the main environmental problems, such as CO_2 emissions, do not (yet) follow this suggested development. The current analysis explores the potential functional forms of the EKC for passenger transport and CO_2 emissions from private car use by augmenting the income determination of these with three variables: dispersion of the population across Austria (with a higher concentration leading to less passenger transport); bringing in a supply-side argument relating to the length of the road network (with extensions to the road network leading to an increase in passenger-kilometres); and including the price level of car use in terms of gasoline prices.

We applied different estimation approaches, the one that performed best being the estimation based on ML and accounting for heteroscedasticity.

Passenger transport and CO_2 emissions from private car use developed roughly proportionally to GDP for the years since World War II until about 1975. The oil price shock changed car use fundamentally, and led to policies aimed at increasing energy efficiency. However, this change, observable as a relative decoupling of car use from GDP, was temporary. After a number of years, car use again increased proportional to GDP, even with an upward trend. In order to account for this structural break in the time series, we tested several potential equations. For instance, one equation was estimated that included a dummy variable for the period after 1975, and a combination of the dummy and the income variable. None of these attempts outperformed the current approach.

The results of the empirical estimations can be interpreted in a number of ways. First, the study is a clear corroboration of an N-shaped environmental Kuznets curve during the observation period from 1950 to 2006, for both passenger transport and private car use, and also for CO_2 emissions from private cars. This result does not promote optimism in terms of climate change policies aimed at a substantial reduction in CO_2 emissions because the driving force of income with respect to car use is very strong. Second, while income seems to be the main driver, technological developments fall significantly short of reducing car use and CO_2 emissions. CO_2 emissions in Austria did not decrease per kilometre driven because diesel cars purchased between 2000 and 2006 emitted about 160 grams of CO_2 per kilometre, with even a small upward trend. This means that new technologies are being used to increase the power of cars rather than to make cars more fuel efficient in terms of fuel consumption per kilometre. These results are in line with Tapio (2005), who provides evidence of a nullification of technological progress based on increases in the weight and engine power of cars.

In order to discuss the policy implications of the empirical evidence for mobility demand, we employed a scenario analysis to forecast passenger mobility to 2020. Given the current trends in incomes, and assuming the empirically indicated functional form of the mobility/emissions–income relationship, the scenarios show that even in the case of a complete halting of further road construction, and continuously increasing fuel prices (e.g. by means of annual increases in fuel taxes), passenger transport and private car use will increase by 10 to 15 per cent to 2020. CO_2 emissions can be stabilized only roughly, indicating that it might not be possible to reduce personal mobility as long as new technological progress and smaller cars are not used to dampen CO_2 emissions.

The N-shaped EKC of car use in Austria points to important policy conclusions. On the one hand, the estimations indicate that there is no automatism in reducing car use and CO_2 emissions and increasing income. It seems that there is no satiation point (or income turning point) in sight, at which passenger transport or CO_2 emissions will decrease again. Furthermore, the results suggest that the transport policies pursued so far have had practically no effect on reducing car use and emissions. On the other hand, the only structural break in the time series, due to the effect of the oil price shock, led to a significant, but only temporary stabilization. Taking these two points together, on the basis of this analysis, we can argue that it should be mandatory for policies aimed at combating climate change to increase fuel taxes dramatically, and to cap emissions by means of strict regulatory frameworks, probably complemented by a trading system. Furthermore, it seems that huge public investments in the railroad system have had no dampening effect on private car use because the incentive to use public transportation has been cannibalized by extensions to the road system and low gasoline prices.[6] Finally, the results of this analysis also suggest that the standard economic policy model of increasing income almost invariably leading to more passenger transport and car use (and therefore emissions), needs to be adapted to a development concept that focuses on a fundamental change in the use of resources, and reopens the debate on significantly altering the current focus of private households' behaviour in terms of sustainability (e.g. consumption style, life style).

Acknowledgements

Parts of this research are based on a research project, Environmental quality and economic growth, commissioned by the Austrian Federal Ministry of the Environment. I am grateful to the Austrian Environmental Protection Agency (Umweltbundesamt) for providing the underlying environmental data.

Notes

1 We ignore other modes of transportation such as walking, bicycles, motorcycles, ships and domestic and international flights, this last having shown the highest growth rates over the last few years.
2 Currently, consistent data are available for the period 1990 to 2006. However, earlier statistics show that GHG emissions increased between 1950 and 1990 at a much faster rate.
3 For reasons of space we do not reproduce the details of the estimations which are available upon request from the author.
4 We would achieve a different result if we added the last years of constantly and significantly increasing fuel prices to the time series. However, we only have complete data up to 2006. Therefore, the most recent years are not included in the analysis.
5 For all results on cointegration, detailed estimations are available from the author.
6 Railroad investments are also criticized from the viewpoint of factor productivity for the Austrian economy (Puwein 2007).

References

Auci, A. and Becchetti, L. (2006) 'The instability of the adjusted and unadjusted environmental Kuznets curve', *Ecological Economics*, 60: 282–98.

Bartz, S. and Kelly, D.L. (2008) 'Economic growth and the environment: Theory and facts', *Resource and Energy Economics*, 30: 115–49.

Clark, S.D. (2007) 'Estimating local car ownership models', *Journal of Transport Geography*, 15: 184–97.

Copeland, B.R. and Taylor, M.S. (2004) 'Trade, growth, and the environment', *Journal of Economic Literature*, 42: 7–71.

Dargay, J. (2001) 'The effect of income on car ownership: Evidence of asymmetry', *Transportation Research Part A*, 35: 807–21.

Dargay, J. (2007) 'The effect of prices and income on car travel in the UK', *Transportation Research Part A*, 41: 949–60.

Dargay, J. and Gately, D. (1997) 'Vehicle ownership to 2015: implications for energy use and emissions', *Energy Policy*, 25: 1121–7.

Dasputa, S., Laplante, B., Wang, H. and Wheeler, D. (2002) 'Confronting the environmental Kuznets curve', *Journal of Economic Perspectives*, 16: 147–68.

Dinda, S. (2004) 'Environmental Kuznets curve hypothesis: A survey', *Ecological Economics*, 49: 431–55.

Dinda, S. (2005) 'A theoretical basis for the environmental Kuznets curve', *Ecological Economics*, 53: 403–13.

Friedl, B. and Getzner, M. (2003) 'Determinants of CO_2 emissions in a small open economy', *Ecological Economics*, 45: 133–48.

Galeotti, M., Lanza, A. and Pauli, F. (2006) 'Reassessing the environmental Kuznets curve for CO_2 emissions: A robustness exercise', *Ecological Economics*, 57: 151–63.

Getzner, M. (2008) 'Exploring the determinants of freight transport in Austria', mimeo, Department of Economics, Klagenfurt University.

Hanley, N., McGregor, P.G., Swales, J.K. and Turner, K. (2008) 'Do increases in energy efficiency improve environmental quality and sustainability?' *Ecological Economics* 16(3): 692–709.

Johansen, S. and Juselius, K. (1990) 'Maximum likelihood estimation and inferences on cointegration – with applications to the demand for money', *Oxford Bulletin of Economics and Statistics*, 52: 169–210.

Kwon, T.-H. (2005) 'Decomposition of factors determining the trend of CO_2 emissions from car travel in Great Britain (1970–2000)', *Ecological Economics*, 53: 261–75.

Löfvenhaft, K., Runborg, S. and Sjögren-Gulve, P. (2004) 'Biotope patterns and amphibian distribution as assessment tools in urban landscape planning', *Landscape and Urban Planning*, 68: 403–27.

Lucas, R. (1976) 'Econometric policy evaluation: A critique', *Carnegie-Rochester Conference Series on Public Policy*, 1: 19–46.

Puwein, W. (2007) 'Gesamtwirtschaftliche Aspekte von Investitionen in die Verkehrsinfrastruktur', *WIFO Monatsberichte*, 9: 747–59.

Raux, C. and Marlot, G. (2005) 'A system of tradable CO_2 permits applied to fuel consumption by motorists', *Transport Policy*, 12: 255–65.

Ruzzenenti, F. and Basosi, R. (2008) 'The role of the power/efficiency misconception in the rebound effect's size debate: Does efficiency actually lead to a power enhancement?' *Energy Policy* 36(9): 3626–32.

Spielmann, M., de Haan, P. and Scholz, R.W. (2008) 'Environmental rebound effects of high-speed transport technologies: A case study of climate change rebound effects of a future underground maglev train system', *Journal of Cleaner Production*, 16: 1388–98.

Stern, D.I. (2004) 'The rise and fall of the environmental Kuznets curve', *World Development*, 32: 1419–39.

Storchmann, K. (2005) 'Long-run gasoline demand for passenger cars: the role of income distribution', *Energy Economics*, 27: 25–58.

Tapio, P. (2005) 'Towards a theory of decoupling: degrees of decoupling in the EU and the case of road traffic in Finland between 1970 and 2001', *Transport Policy*, 12: 137–51.

Umweltbundesamt (2008) Verkehr – Auswirkungen – Schadstoffe. http://www.umweltbundesamt.at/umweltschutz/verkehr/auswirkungen_verkehr/verk_schadstoffe/, Umweltbundesamt (UBA), Vienna, accessed 24 July 2008.

Wadud, Z., Noland, R.B. and Graham, D.J. (2008) Equity analysis of personal tradable carbon permits for the road transport sector', *Environmental Science & Policy* 11(6): 533–44.

Wohlgemuth, N. (1997) 'World transport energy demand modelling'. *Energy Policy*, 25: 1109–19.

Part III

Economic growth, environmental degradation and policies

10 Reconsidering the environmental Kuznets curve hypothesis

The trade-off between environment and welfare

Nicola Cantore and Maurizio Canavari

Introduction

Much of the existing environmental Kuznets curve (EKC) literature focuses on empirical evidence on the relationship over time between economic activity and pollution, and tends to ignore the welfare implications of different growth paths.

The appeal of the EKC concept lies in the fact that if confirmed by the data it offers a solution to the growth/pollution trade-off, and demonstrates that increasing GDP is the best way to tackle this environmental problem (Grossman 1995). The shortcoming of this concept is that it incorporates only two policy dimensions, the economy and the environment, but does not consider other relevant policy targets such as welfare.

In our view, the bell-shaped pollution–income relationship (PIR) could be hiding a level of welfare that is lower than in other less environment-friendly scenarios. The complexity of these economic relations implies that the shape of the PIR should not be the only factor considered in deciding on the merits of a particular scenario: there are other indicators that need to be taken into account for the achievement of policy targets.

In this chapter we verify the shape of the PIR and the welfare level for different policy scenarios. The novelty of our contribution is that it deals with a topic often neglected by the scientific EKC literature: the trade-off that might arise between environmental improvement and welfare reduction.

Two points are important here. The first is that the traditional EKC concept implies that if the PIR is bell shaped, growth will be the best solution to the environmental problem, and there will be no need for policies to reduce global warming (Arrow *et al.* 1995). For the most part, in the EKC literature checking for the presence/absence of a turning point is seen as the best basis for decisions about pollution reduction. The existence of an EKC implies that a turning point can be achieved endogenously by the economic system, without the need for public policy actions.

However, a few studies deal with the EKC hypothesis by analysing the impact of different environmental policies on the PIR. In this interpretation, a turning point can be reached by the implementation of environmental policy rather than by endogenous mechanisms of the economic system.[1] The focus of this interpretation is the simple confirmation of the presence of a turning point rather than how it is reached. In this chapter, we focus on this latter type of EKC analysis, and investigate the impact of different emissions on policies for stabilizing the PIR.

We also investigate the level of welfare for different scenarios using tools derived from the social choice literature. In this literature, level of welfare depends also on gross domestic product (GDP) and how it is shared among countries in terms of inequalities. Since, unlike some of the previous studies, inequality in income distribution is a key variable in our study, we use a multi-agent model RICE99 (Nordhaus and Boyer 1999). Nordhaus and Boyer's RICE model is a regional dynamic general equilibrium model for studying the economic aspects of climate change. This model is based on a single sector optimal growth model suitably extended to incorporate the interactions between economic activities and climate, in which the world is divided into eight macro regions: United States of America (USA), Western Europe (Europe), Other High Income countries (OHI), Eastern European countries (EE), Middle Income countries (MI), Lower Middle Income countries (LMI), China (CHN), and Low Income countries (LI).

Within each region a central planner chooses the optimal paths of fixed investment and carbon energy input, to maximize the present value of per capita consumption as in the usual Ramsey–Solow–Koopmans framework. Environmental damage is introduced as an output reducing the effect of global warming, which is expressed as the willingness of each region to pay to avoid future climate change impacts.

This chapter makes two contributions. Using RICE99 we investigate: (i) whether international policies can generate a turning point in the relationship between income and pollution as postulated by the EKC hypothesis; and (ii) their welfare effects on the basis of New Welfare Economics (NWE) and non-NWE approaches.

The scenarios

We run four scenarios for RICE99 from 1995 to 2085: Business as usual (BAU), Kyoto forever scenario (Kyoto), Global Kyoto scenario (GK), and Global Kyoto scenario 2 (GK2). The BAU scenario implies no policy regulation for emissions. In this scenario there are no limits on economic growth for any of the world regions. The non-cooperative game induces every social planner to foster economic development by increasing emissions outputs, which influences temperature and climate change.

The Kyoto forever scenario implies a constraint over time on the emissions for Annex I[2] regions (USA, OHI, Europe, EE) imposed by the Kyoto Protocol. We also include in the models an emissions trading market[3] because we assume that future emissions caps will be implemented by an efficient mechanism to minimize global abatement costs. The implementation of an emissions permits market does not change our results for the relationship between income and pollution because it does not affect global emissions reductions over time when compared with a command-control emissions policy. However, it influences the redistribution of welfare costs among countries, which depends on the structure of regional marginal abatement costs and the magnitude of the agents' emissions caps.

The GK scenario implies an emissions constraint for every world region including developing countries (MI, LMI, CHN and LI). Given that the Kyoto Protocol does not currently include an emissions constraint for developing countries, we assume that the emissions constraint for each developing region is the level of emissions registered in that region in 2015. Moreover, we include a global market for emissions trading.

The GK2 scenario assumes a Kyoto-forever constraint for developed countries as in the GK scenario. After 2025, the emissions constraint over time for developing countries is 80 per cent of the previous BAU period emissions. In this case the constraint is less rigid, with no fixed cap for the poor regions.

The four scenarios are summarized in Table 10.1. There is no reason to believe that these scenarios are the most plausible; they were chosen based on their appropriateness for answering the questions posed in this work. Global warming has reached worrying proportions and has provoked lively debate over emissions-stabilizing policies. In this chapter we investigate first whether a BAU scenario generates a turning point. If no turning point is reached and environmental policies are needed, then we need to investigate the role played by international emissions-stabilizing policies.

We examine whether a constraint on the emissions from Annex I regions (Kyoto forever scenario) would be a policy option that achieves a turning point, or whether it will be necessary to impose emissions constraints on the developing countries (GK, GK2 scenario). We assume two possible emissions-reductions policies for the developing countries to check how the relationship between income and pollution would be affected by different emissions constraints for the poorest regions.

It should be pointed out that our Kyoto forever scenario involves the USA, which in 2009 had not yet ratified the Kyoto Protocol. Thus, although a Kyoto forever scenario is not presently realistic, we impose an emissions cap for the

Table 10.1 Description of the scenarios

Scenario	*Description*
BAU	No policy
Kyoto	Since 2015 Kyoto forever emission constraint for Annex I regions and market of pollution permits
Global Kyoto	Since 2015 Kyoto forever emissions constraint for Annex I regions. Developing regions are obliged to maintain the level of emissions as in 2015. Market of pollution permits
Global Kyoto 2	Since 2015 Kyoto forever emissions constraint for Annex I regions. Since 2025 developing regions are obliged to maintain 80% of the previous period BAU emissions. Market of pollution permits

USA because our interest is in analysing the impact of international policies involving the widest and most significant developed countries coalition.[4] We investigate whether emissions-reducing efforts of the Annex I regions overall, as imposed by the Kyoto Protocol, could generate a turning point in the relationship between income and pollution, and their impact on welfare. For the other three scenarios, we study whether the existence of a turning point is compatible with welfare improvements in emissions-stabilizing policies.

Simulation results

To analyse the relationship between income and air pollution we focus on the relationship between global per capita gross domestic product (GDP) and per capita emissions. A per capita version of the environmental indicator is often used in empirical studies of EKC (Panayatou 1993; Selden and Song 1994).

Our choice of the global scale was made to avoid discrepancies between the EKC for individual countries and a world EKC (Dinda 2004). According to the EKC hypothesis, improvements in environmental quality are not attainable for the majority of the world's population, whose living standards are substantially below the estimated turning points (Stern *et al.* 1996). We need to investigate the relationship between income and pollution for the world. The time period we consider is 1995–2085.

Figure 10.1 shows that the BAU scenario predicts an increasing relationship between per capita income and per capita emissions, and in the simulations driven by the RICE99 model the relationship is shown to be first decreasing and then continuously increasing without reaching a turning point. If there is no evidence of an EKC and there is an increasing relationship between income and pollution, this indicates the need for environmental policies aimed at reducing emissions levels.

Figure 10.1 depicts the environmental consequences deriving from the design of international environmental agreements. The plot shows that only GK generates

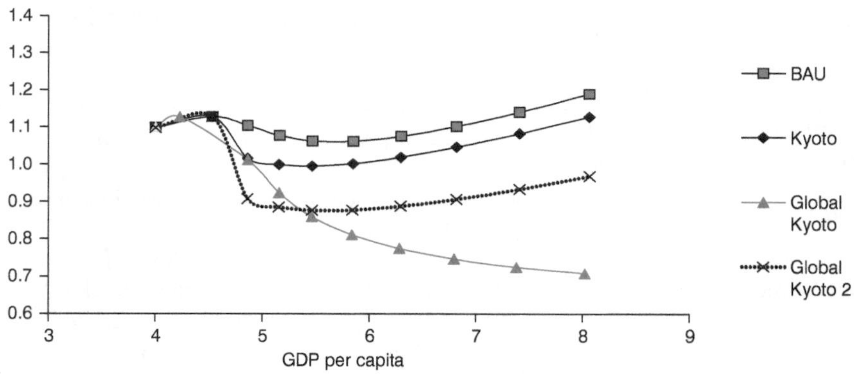

Figure 10.1 World GDP per capita (trillions of 1990 US$/billions of population) vs world emissions per capita (gigatons/ billions of population), 1995–2085

a decreasing PIR and a bell-shaped relationship between income and pollution, as estimated by the RICE99 model. Thus, only if the Kyoto Protocol constraints can be extended to developing countries we can be optimistic about global environmental sustainability. In contrast, the GK2 scenario shows an increasing relationship between income and pollution based on the less strict emissions constraint for the developing countries. We want to establish the consequences in terms of variation in welfare, deriving from the different emissions stabilizing policy designs.

Welfare definitions

Our first indicator is a traditional measure of welfare, which does not include equality criteria and represents a metric measure based on economic values. Buchner and Galeotti (2004) implement a welfare analysis using RICE models, considering the present discounted value of consumption. Here, we consider the present discounted value of world consumption[5] over time. The measure of welfare is expressed by the following equation:

$$W(t) = \sum_{t=1}^{T} \beta(t)^{t-1} C(t) \tag{10.1}$$

where $C(t)$ is world consumption in each period t, and $\beta(t)$ is the discount factor. We consider it reasonable to use consumption rather than income as the indicator of welfare because consumption is the endogenous variable of the utility function in the RICE99 model. This simple measure of welfare, of course, does not capture the distribution of consumption among regions over time.

The results deriving from this Benthamian measure of welfare can be compared with measures that include equality principles. We choose three indices of welfare based on different approaches to dealing with redistribution.[6] The first is Sen's index, which derives from the contributions on equality. Atkinson (1970) represents the cornerstone of this strand of the literature. He points to the role of income distribution in determining welfare rankings. Atkinson proves that assuming a diminishing marginal utility of income, allows a Lorenz curve[7] ordering to rank alternative social states with the same average income. Shorrocks (1983) extends Atkinson's formulation and introduces the concept of generalized Lorenz dominance that is useful when Lorenz curves intersect. If the Lorenz curve of a distribution is $L(p)$ and the mean income of the distribution is μ, then the generalized Lorenz curve of this distribution is $\mu L(p)$. According to Shorrocks, if the generalized Lorenz curve of one state lies above the curve for another, the social welfare corresponding to the former is assumed to be superior to the latter.[8]

Both Lorenz dominance and generalized Lorenz dominance provide only a partial ordering of social welfare over time. For a complete ordering we introduce a cardinal social welfare function (SWF), which provides numerical values for all possible social states. As the Gini index (G) is defined as twice the area between the Lorenz curve and the egalitarian diagonal line, and is included within the range [0, 1], then a cardinalization of the SWF can be obtained by calculating (1 –

G), which is twice the area below the generalized Lorenz curve. On the basis of a set of axioms, Sen (1974) proposes a specific form of the SWF which is:

$$W(t) = \mu(t) \cdot (1 - G(t)) \tag{10.2}$$

In our analysis μ is world consumption per capita and G is the Gini consumption concentration index among regions.

Another important SWF, which includes equality issues, derives from the wide literature that criticizes the use of GDP as the only factor influencing well-being. England (1998) surveys a number of quantitative measures of well-being that have been proposed as complements to or substitutes for GDP.

Based on these alternative approaches, the United Nations Development Programme (UNDP) developed a Human Development Index (HDI) to represent a measure of national well-being. One of the components of the HDI is real per capita GDP[9] and, since 1991, an Atkinson-style utility function is used to transform national income level into a measure of national social welfare U for each region n:

$$U(n,t) = \frac{y(n,t)^{(1-\varepsilon)}}{(1-\varepsilon)} \tag{10.3}$$

where y is a country's per capita income. When y^* is global per capita income, for nations with a per capita GDP above the global average $\varepsilon = \alpha/(\alpha + 1)$, where α is a non-negative integer and $\alpha y^* \leq y \leq (\alpha + 1)y^*$. Alternatively, $\varepsilon = 0$ for nations with a per capita GDP below the global average; in this case $U(n, t) = y(n, t)$.

Although UNDP SWF is supposed to transform national income levels, with RICE99 we deem it more reasonable to use consumption. Moreover, given that $U(n, t)$ is the utility function of each region n, we assume only that the global SWF is an additional and separable function of the utilities (Atkinson 1970) enjoyed by each region. Thus, the global SWF function becomes:

$$W(t) = \sum\nolimits_n U(n,t) \qquad \text{where } U(n,t) = \frac{c^{1-\varepsilon}}{1-\varepsilon} \tag{10.4}$$

where c is consumption per capita. Combined with equations (10.1), (10.2) and (10.4), we use a further SWF to assess welfare in the different environmental policies scenarios. As already mentioned, Sen (1999) criticizes the NWE approach and considers that in the social choice process interpersonal comparisons based on deprivation and mental states are both crucial. Several economists share Sen's view (Kahneman *et al.* 1997; Gowdy 2005).

Although in many environmental models the utility function is characterized by psychological assumptions about human behaviour (monotonicity, non-satiation), economists generally prefer monetary measures of well-being to extract useful welfare implications. As an indicator of well-being in RICE99, we consider again an additive and separable SWF depending on the utility function of each region as expressed in RICE99. Therefore:

$$W(t) = \sum_n U(n,t) \quad \text{where } U(n,t) = \sum_{t=1}^{T} \beta(t)^{t-1} \left(L(n,t) * \log\left(C(n,t) / l(n,t) \right) \right) \tag{10.5}$$

With this SWF, the transfer of one unit of consumption per capita from the wealthiest countries to the developing regions is welfare improving.

Application of these welfare functions in RICE99 implies the existence of a world social planner (WSP) aimed at assessing the results deriving from different scenarios and from the maximization process in each region. The WSP considers the data deriving from RICE99 as representing the future real world, and maximizes an additive and separable SWF depending on the utility function of each individual. As the RICE99 model is based on a regional framework, and since disaggregated data for each variable are not available, in our analysis we assume implicitly that consumption in each region is distributed equally among the population. In other words, we investigate a between-group rather than a within-group approach to inequality.

Equations (10.1), (10.2), (10.4) and (10.5) assess international emissions-reducing policies. We are interested in evaluating whether different notions of SWF based on different approaches, produce the same conclusions.

Welfare results

First we present the results for the traditional measure of welfare. Table 10.2 shows that, in this case the BAU scenario is the best over time because it produces the highest levels of discounted consumption.[10]

This result is hardly surprising. Emissions constraints are costly in terms of consumption. In the Kyoto forever scenario, developed countries must manage to comply with the binding emissions constraints and are committed to reducing their levels of economic activity and, consequently, their levels of consumption, while developing countries enjoy the lower levels of environmental damage induced by the Kyoto agreement. However, the losses suffered by the developed countries outweigh the benefits to developing countries and levels of global consumption decrease.

In the GK and GK2 scenarios, welfare losses are even higher than in a Kyoto scenario. Emissions constraints strongly penalize growth and consumption in the poor regions. World consumption is lower because both developed and developing countries are subject to the emissions constraints and are obliged to decrease their levels of economic activity.

The results change when we include equality criteria in our analysis. Table 10.3 shows the results for Sen's SWF. From 2015 a Kyoto forever scenario always shows welfare gains. Scenarios that include the developing countries are the worst over time in terms of welfare.

Table 10.4 shows this more clearly by presenting per capita consumption and the value associated with the Gini coefficient for each scenario.

Over time, the optimum scenario in terms of per capita consumption is BAU (except for 2085, when less environmental damage generates higher per capita

Table 10.2 A traditional measure of welfare: % differences of environmental policies vs BAU scenario (1995–2085)

Year	*Kyoto vs BAU*		*GK vs BAU*		*GK2 vs BAU*	
1995	0.004		0.005		0.004	
2005	0.006		0.007		0.007	
2015	–0.014		–0.001		–0.013	
2025	–0.024		–0.017		–0.052	
2035	–0.028		–0.034		–0.074	
2045	–0.029		–0.053		–0.085	
2055	–0.029		–0.073		–0.091	
2065	–0.028		–0.092		–0.093	
2075	–0.027		–0.111		–0.093	
2085	–0.025		–0.129		–0.091	
Year	*Annex I*	*Non Annex I*	*Annex I*	*Non Annex I*	*Annex I*	*Non Annex I*
1995	0.005	0.000	0.005	0.003	0.005	0.003
2005	0.008	0.000	0.008	0.005	0.008	0.005
2015	–0.021	0.000	–0.009	0.020	–0.020	–0.076
2025	–0.037	0.000	–0.024	–0.009	–0.035	–0.115
2035	–0.045	0.002	–0.033	–0.067	–0.042	–0.134
2045	–0.048	0.006	–0.037	–0.141	–0.044	–0.140
2055	–0.050	0.011	–0.037	–0.223	–0.044	–0.139
2065	–0.051	0.016	–0.033	–0.310	–0.042	–0.134
2075	–0.052	0.022	–0.026	–0.398	–0.040	–0.126
2085	–0.052	0.027	–0.017	–0.484	–0.036	–0.116

Table 10.3 The Sen's SWF. Time period: 1995–2085. % differences of environmental policies vs BAU scenario

Year	*Kyoto vs BAU*	*GK vs BAU*	*GK2 vs BAU*
1995	0.004	0.005	0.004
2005	0.007	0.011	0.009
2015	0.018	0.019	0.023
2025	0.014	–0.067	–0.336
2035	0.023	–0.204	–0.324
2045	0.040	–0.370	–0.283
2055	0.058	–0.567	–0.236
2065	0.071	–0.795	–0.187
2075	0.080	–1.045	–0.133
2085	0.088	–1.308	–0.071

consumption). The Kyoto forever scenario produces the best results in terms of equality. An emissions constraint involving only the developed regions reduces growth in the wealthy countries and, consequently, world growth, but increases equality over time. The Kyoto forever scenario produces a trade-off: lower consumption per capita, but greater equality than a BAU scenario. According to Sen's SWF the trade-off is resolved in favour of the Kyoto forever scenario. In the GK and GK2 scenarios, imposing an emission constraint for the developing countries results in lower per capita consumption and less equality than in a BAU or a Kyoto scenario. Limits on poor regions reduce world consumption growth and generate greater inequality.

Only the Kyoto forever scenario always shows welfare gains. These results suggest a strong preference for an emissions stabilizing policy involving only the developed countries.

Table 10.5, and considering the UNDP SWF, shows that these results are quite similar. The Gini concentration index in Equation (10.4) does not play a role, but consumption per capita for each region is weighted according to equality criteria. The SWF attributes a higher weight to increases in consumption for the poor regions.

Given the framework of Equation (10.4), in a Kyoto forever scenario, the decrease in per capita consumption for the developed countries subject to an emissions constraint, has a lower impact than the gains deriving from a lower level of environmental damage in the developing regions. Thus, the Kyoto forever scenario provides the best welfare performance compared with BAU.

Again, GK and GK2 show high levels of welfare losses deriving from the high per capita consumption reductions in developing countries generated by the emissions constraints. However, as in the previous case the welfare loss seems to increase over time, with a stricter emissions constraint in the GK scenario.

Finally, we analyse social welfare based on a subjective definition of welfare. Table 10.6 shows that once again the Kyoto forever scenario is welfare improving over time and the least favourable scenarios are those involving the developing

Table 10.4 The Sen's SWF decomposition by per capita consumption (PCC, trillions of 1990 US$/billions of population) and the Gini index (1995–2085)

	BAU				*Kyoto*			
Year	*PCC*	*Gini*	*1-GINI*	*Utility*	*PCC*	*Gini*	*1-GINI*	*Utility*
1995	2.979	0.643	0.357	1.062	2.979	0.643	0.357	1.062
2005	3.432	0.610	0.390	1.338	3.432	0.610	0.390	1.338
2015	3.710	0.582	0.417	1.548	3.707	0.582	0.418	1.548
2025	3.948	0.556	0.443	1.751	3.945	0.556	0.444	1.751
2035	4.180	0.530	0.470	1.963	4.177	0.530	0.470	1.963
2045	4.463	0.506	0.493	2.204	4.461	0.505	0.494	2.205
2055	4.800	0.483	0.516	2.478	4.799	0.483	0.517	2.480
2065	5.190	0.463	0.537	2.787	5.189	0.462	0.537	2.789
2075	5.630	0.444	0.556	3.131	5.630	0.443	0.556	3.133
2085	6.122	0.426	0.574	3.512	6.124	0.426	0.574	3.515

	GK				*GK2*			
Year	*PCC*	*GINI*	*1-GINI*	*Utility*	*PCC*	*GINI*	*1-GINI*	*Utility*
1995	2.979	0.643	0.357	1.062	2.979	0.643	0.357	1.062
2005	3.432	0.610	0.390	1.338	3.432	0.610	0.390	1.338
2015	3.709	0.583	0.417	1.548	3.707	0.582	0.418	1.548
2025	3.945	0.556	0.444	1.750	3.939	0.557	0.443	1.745
2035	4.173	0.531	0.469	1.959	4.170	0.531	0.469	1.957
2045	4.452	0.507	0.493	2.196	4.454	0.507	0.493	2.198
2055	4.784	0.485	0.515	2.464	4.792	0.484	0.516	2.472
2065	5.166	0.465	0.535	2.765	5.183	0.463	0.537	2.781
2075	5.598	0.447	0.553	3.098	5.625	0.444	0.556	3.126
2085	6.080	0.430	0.570	3.466	6.121	0.427	0.573	3.509

Table 10.5 The UNDP SWF. Time period: 1995–2085. % differences of environmental policies vs BAU scenario

Year	*Kyoto vs BAU*	*GK vs BAU*	*GK2 vs BAU*
1995	0.002	0.002	0.002
2005	0.004	0.004	0.004
2015	0.021	0.002	0.022
2025	0.020	–0.004	–0.027
2035	0.018	–0.014	–0.029
2045	0.017	–0.028	–0.029
2055	0.017	–0.048	–0.027
2065	0.018	–0.077	–0.024
2075	0.019	–0.114	–0.019
2085	0.019	–0.162	–0.012

Table 10.6 A "subjective" SWF. Time period: 1995–2085. % differences of environmental policies vs BAU scenario

Year	*Kyoto vs BAU*	*GK vs BAU*	*GK2 vs BAU*
1995	0.000	0.001	0.001
2005	0.001	0.002	0.001
2015	0.002	0.003	0.002
2025	0.002	0.000	–0.012
2035	0.002	–0.006	–0.019
2045	0.003	–0.013	–0.024
2055	0.003	–0.021	–0.026
2065	0.004	–0.029	–0.027
2075	0.004	–0.038	–0.027
2085	0.004	–0.046	–0.027

countries (GK and GK2). However, when the emissions constraint for developing countries is less severe, as in the GK2 scenario, welfare losses are lower. The utility function of each region is less equality oriented in the subjective welfare index than in the UNDP index. Whereas in the subjective index the logarithmic form of the utility function is the same for all regions, in the UNDP index the elasticity of the marginal utility of consumption ε is higher for wealthy regions and, consequently, the weight for developing regions' consumption losses is higher. A less equality oriented utility function, the presence of a discount rate, and the fact that we sum population weighed against sums of consumption per capita over time, do not qualitatively change the results compared with those based on the UNDP index.

The above results explain quite clearly the reluctance of the European Union (EU) to involve the developing countries in an international emissions-stabilizing agreement: a small environmental improvement without negative global welfare implications, as in the Kyoto Protocol design, is preferable to policies aimed at achieving a turning point in the PIR, which would incur high costs in terms of well-being, as in the GK scenario.

Of course, the EU positions in relation to ratification of the Kyoto Protocol cannot be interpreted only on the basis of altruistic considerations. The literature highlights that the soft position taken by the EU is determined mainly by the lower costs of abatement compared with those faced by the USA, for Kyoto compliance (Nordhaus and Boyer 1999).

However, our aim is to underline a crucial consideration for policy that when a third dimension (e.g., welfare) is introduced into the EKC analysis, the existence of a turning point is not crucial for the extraction of useful environmental policy implications. In our simulations, the Kyoto forever scenario results in reduced emissions when compared with a BAU scenario, and is welfare improving on the basis of measures that include economic and equality components. The implementation of win–win policies, as in the Kyoto forever scenario, which does not generate a turning point in the relationship between income and pollution, might be preferable to strong welfare-reducing policies, as in the GK scenario.

Conclusions

This chapter investigated four scenarios involving different hypotheses related to the participation of different regions in international emissions-stabilizing programmes. Following a multidimensional approach to sustainability, we analysed the economic, social and environmental impacts of each scenario. Specifically, we studied the relationship between income and pollution. Also, using non-NWE approaches, we analysed the level of welfare associated with each scenario. We found that the only scenario implying a contextual welfare and environmental improvement is Kyoto forever, which does not generate a turning point in the PIR. Emissions constraints for developing countries are strongly welfare reducing. There are two important policy implications from these results.

First, although results should be taken with some caution because they depend on the calibration of the model and various assumptions that we discussed above, the choice of the welfare measure used to analyse policies is crucial. When we consider welfare measures that incorporate an equality criterion, the implementation over time of the Kyoto Protocol is preferable to a no-policy scenario. When we use a traditional measure of welfare based on consumption, the results are reversed. The welfare measure is a tool that should be exploited by policy makers with caution. The most appropriate welfare measure will be chosen on the basis of the policy targets.

One drawback to our analysis is that it is based on global measures of welfare. The underlying hypothesis is that a World Agency would be interested in accounting for world welfare levels in different scenarios, and choosing the best option on the basis of general interest. This could be unrealistic; economic theory explains that in the case of issues related to global warming, the most realistic view is to consider maximizing agents that behave on the basis of their own self-interest. However, recent policies related to debt cancellation and ratification of the Kyoto Protocol by a large majority of Annex I countries, exemplify that, in future, ethics and social justice may be more likely to drive international policies in a context of achieving global sustainability. The analysis in this chapter adopts this alternative view.

Second, we analysed the role played by the existence of a turning point in the relationship between income and pollution as a criterion in the formulation of environmental policies. The EKC hypothesis postulates that if there is a bell-shaped relationship between income and pollution, fostering economic growth is the best policy for tackling pollution. The existence of a turning point has been considered a crucial factor in environmental policy setting. The theoretical and empirical literature focuses mainly on the shape of the relationship between income and pollution, an approach that we criticize here. We show that the existence of a turning point could be an unsuitable social choice criterion. The EKC growth path resolves the traditional trade-off between income and the environment, but it could be strongly welfare reducing. Empirical observation of the relationship between income and pollution provides useful evidence for policy makers engaged in setting environmental policy, but cannot be assumed to be the only relevant criterion. A search for *win–win* policies generating economic, social and environmental improvements although not generating a turning point, would seem to be a more appropriate method for social choice.

However, if *win–win* policies are not possible, social choice theory should be applied to resolving the trade-offs. Following Atkinson's (1970) findings, the literature on equality shows increased opportunities for achieving social choice through management of the trade-off between the economic and social variables. Further efforts should be made to build SWF to manage the trade-offs related to all dimensions of sustainability: economic, environmental and social. The policy prescription deriving from the EKC hypothesis seems to be not sufficient to derive useful insights when several policy targets are involved. When other policy targets than environment and income only are involved, using the EKC alone may result in inappropriate policies.

To conclude then, rather than focusing on the existence of a turning point in the relationship between income and pollution, future research should investigate multidimensional paths using relevant variables. This study constitutes a first contribution in this direction.

Acknowledgements

We are grateful to Vania Paccagnan, John Hutton, Riccardo Scarpa and Charles Perrings for useful suggestions and comments. The authors acknowledge the support of the project Modelli matematici per le decisioni economico – finanziarie – attuariali, D.1 Sedi Padane – Anno 2008. Maurizio Canavari wrote the introduction and conclusions, Nicola Cantore the remaining paragraphs.

Notes

1 E.g., Lantz and Feng (2005) discuss the effect of Canada's signing the Kyoto Protocol in terms of generating a turning point in the PIR.
2 The Kyoto Protocol designs as Annex I countries those which committed themselves as a group to reduce their emissions of greenhouse gases.
3 We adopt the same open loop Nash equilibrium technique as in Bosetti *et al.* (2006).
4 In any case, the inclusion of the USA does not change the conclusions reached in our analysis. The results for the model without Kyoto protocol constraints for the USA, are available upon request to the authors.
5 The correct measure of welfare would be consumption per capita rather than just consumption. However, for welfare indices that do not include equality aspects, this measure is reliable because, in the model, population is an exogenous variable that does not change across scenarios.
6 There is a wide set of possible social welfare functions that we do not consider in this chapter, e.g. the maximin welfare function and the Yitzhaki (1983) abbreviated welfare function. In this chapter, in order to produce interesting comparisons, we use a restricted set of indices with different foundations.
7 The Lorenz curve for income distribution plots the cumulative share of population ranked according to the per capita income vs cumulative share of total income held by that population. It describes graphically concentration vs equal distribution of income among the population.
8 Other theoretical contributions can provide tools to rank alternative social states when generalized Lorenz curves intersect (Atkinson and Bourgignon 1987; Dardanoni and Lambert 1988).
9 The Human Development Index also focuses on variables such as longevity and access to education. The algebraic transformation of consumption levels we use in this chapter was adopted until 1999 and then replaced by a formula based on natural logarithms.
10 Of course, our discussion does not consider resilience, stability or catastrophes deriving from climate change, which are beyond the scope of the investigation in this chapter.

References

Arrow, K., Bolin, B., Costanza, R., Folke, C., Holling, C.S., Janson, B., Levin, S., Maler, K., Perrings, C. and Pimental D. (1995) 'Economic growth, carrying capacity and the environment'. *Science*, 15: 91–5.

Atkinson, A. (1970) 'On the measurement of inequality', *Journal of Economic Theory*, 2: 244–63.

Atkinson, A. and Bourgignon, F. (1987) 'Income distribution and differences in needs' in G.R. Feiwel (ed.), *Arrow and the Foundations of the Theory of Economic Policy,* London: McMillan, ch. 12.

Bosetti, V., Carraro, C. and Galeotti, M. (2006) 'The dynamics of carbon and energy intensity in a model of endogenous technical change', *The Energy Journal,* Special Issue on Endogenous Technological Change and the Economics of Atmospheric Stabilization, 195–206.

Buchner, B. and Galeotti, M. (2004) 'Climate policy and economic growth in developing countries: The impact of Kyoto', *International Journal of Global Enviromental Issues*, 4: 109–38.

Dardanoni, V. and Lambert, P. (1988) 'Welfare rankings of income distributions: A role for the variance and some insights for tax reform', *Social Choice and Welfare*, 5: 1–17.

Dinda, S. (2004) 'Environmental Kuznets curve hypothesis: A survey', *Ecological Economics*, 49: 431–55.

England, R. (1998) 'Measurement of social well-being; alternatives to gross domestic product', *Ecological Economics*, 25: 89–103.

Gowdy, J. (2005) 'Toward a new welfare economic for sustainability', *Ecological Economics*, 53: 211–22.

Grossman, G.M. (1995), 'Pollution and growth, what do we know?' in I. Goldin and L.A. Winters (eds) *The Economics of Sustainable Development*, New York: Cambridge University Press, pp. 19–46.

Kahneman, D., Wakker, P. and Sarin, R. (1997) 'Back to Bentham? Explorations of experienced utility', *Quarterly Journal of Economics,* 112: 375–405.

Lantz, V. and Feng, Q. (2005) 'Assessing income, population and technology impacts on CO_2 emissions in Canada: Where is the EKC?', *Ecological Economics*, 57: 229–38.

Nordhaus, W. and Boyer, J. (1999) Rolling the DICE again: Economic models of global warming, Cambridge, MA: MIT Press.

Panayatou, T. (1993) 'Empirical tests and policy analysis of environmental degradation at different stages of economic development', *Working paper WP 238,* Technology and Employment Programme, International Labour Office, Geneva.

Selden, T. and Song, D. (1994) 'Environmental quality and development: Is there a Kuznets curve for air pollution?', *Journal of Environmental Economics and Management*, 27: 147–62.

Sen, A. (1974) 'Information bases of alternative welfare approaches: Aggregation and income distribution', *Journal of Public Economics,* 3: 387–403.

Sen, A. (1999) 'The possibility of social choice', *American Economic Review*, 89: 349–78.

Shorrocks, A. (1983) 'Ranking income distributions', *Economica*, 50: 1–17.

Stern, D.I., Common, M.S. and Barbier, E.B. (1996) 'Economic growth and environmental degradation: A critique of the environmental Kuznets curve', *World Development,* 24: 1151–60.

Yitzhaki, S. (1983) 'On an extension of the Gini index', *International Economic Review*, 24: 617–28.

11 Environmental degradation, economic growth and welfare

The role of self-protection choices

Angelo Antoci and Simone Borghesi

Introduction

Increasing attention is being paid to the economic consequences of environmental degradation, in terms of both its impact on human health and productivity, and the estimated large abatement costs that will be required to mitigate global environmental problems. However, rather less attention is being given to some of the perverse effects that environmental degradation may generate on economic growth and the well-being of individuals, through the conditioning of their consumption choices. In particular, as some studies point out (Hueting 1980; Antoci and Bartolini 1999, 2004; Bartolini and Bonatti 2002, 2003; Lopez 2003; Antoci *et al.* 2008), environmental depletion may be modifying prevailing consumption patterns, leading individuals to substitute free environmental goods with costly private goods that allow the agents to protect themselves against environmental degradation.[1]

In the industrial economies there is a wide range of private goods and services available to agents, that can be used as instruments of self-protection against environmental degradation. The most typical and more frequently quoted examples include air filters and water treatment plants, mineral water, double-glazing to reduce the noise from urban traffic, medication for pollution-related diseases (e.g. asthma and various skin diseases).

Degradation of coastal areas that are near to urban centres is occurring as a result of overbuilding and environmentally unfriendly activities in surrounding industrial areas, which is inducing urban residents to implement a wide range of self-protection choices. Living near to a coastal area enables enjoyment of environmental goods at low cost. However, degradation of local areas encourages trips to less contaminated areas by car or by boat, or the purchase of expensive holidays in some tropical paradise, thus forcing individuals to pay for a good that once was freely available.

The examples mentioned so far can be classified as environmental self-protection choices. However, this notion has a broader interpretation, which includes the much wider set of consumption choices that are partially (but not exclusively) caused by environmental degradation. Urban life provides several examples of such choices.

The increasing use of home entertainments, for instance, may be due in part to a scarcity of environmental goods in the residential area (e.g. urban parks for children to play in), which means that many individuals spend more time in the home rather than engaging in outdoor activities. The increase in the number of fitness centres and swimming pools also derives in part from the shortage of outdoor spaces for open-air sports, and of clean rivers, seas and lakes for bathing.

Similarly, the choice of using the car in town instead of travelling by bicycle or on foot may be a response to several non-environmental needs (e.g. to move faster and to cover longer distances, to be protected from bad weather, to display one's social status, etc.). However, it may also be determined by ecological reasons: the higher the urban air pollution levels, the greater the incentive to travel by car in order to reduce exposure to the air pollutants being released by the traffic. Paradoxically, this increases urban traffic and air pollution, thus reinforcing the choice to travel by car. Therefore, the production and consumption of private goods that are used to substitute for previously free environmental goods can contribute to further increasing environmental degradation. Air conditioning is another paradigmatic example of this process. Air-conditioning units are used to defend against global warming by cooling the interiors of homes and offices; however, they give off heat, thus generating an increase in external temperatures, which then encourages their use. As these examples show, the choice of private substitutes to defend against environmental degradation can lead to a final outcome in which their use is above socially optimal levels.

Another self-protection choice that can further increase environmental degradation is moving house to live in an area outside a degraded urban centre. This choice, which usually is motivated by both economic (the search for a less expensive house) and environmental (traffic noise, urban air pollution, desire for bigger living space, etc.) reasons, has led to increasing 'urban sprawl' invading the countryside. This phenomenon, ultimately can exacerbate the problems that originally induced the out-of-town flows, such as reduction in green and public areas, traffic congestion and accompanying increased polluting emissions, and length of the time spent commuting to work places. Environmental degradation, therefore, may encourage urban sprawl, which in turn may increase environmental degradation and extend the ecological problems to new residential areas. Thus, agents who move away from towns in search of better living conditions, potentially could become worse off based on higher travel costs and reduced leisure time, and no significant improvement in terms of their surrounding environmental conditions.

As these examples show, although these self-protection choices are perceived as individually rational (i.e., utility maximizing for the agents who make them), ultimately, they may reduce the welfare of the whole population at the aggregate level. The mechanism underlying these perverse effects can be summarized as follows. To defend against environmental degradation, economic agents make self-protection choices through the consumption of private goods that may satisfy the same needs that previously were met by free environmental goods. The production and consumption of these private goods, in turn, may increase the environmental degradation, thus further increasing the incentive to produce and consume private goods as a self-protection device. Thus, the substitution of environmental with private goods may produce a vicious circle, which will establish an undesirable growth path along which economic growth is accompanied by a reduction in individuals' welfare.

This chapter demonstrates the strategic context underlying the mechanism described above based on a very simple analytical model.

Undesirable effects of environmental self-protection choices: a two-island model

To illustrate the substitution mechanism described above we consider a population of individuals who live and work on an island, which we will call island A. To defend against the environmental degradation of island A, each individual has the possibility of buying a boat that can be used to move to an uncontaminated island B. In order to buy the boat, agents have to work and produce more. We assume that the consequent increase in the production activity tends to raise the environmental degradation of island A and, to a lesser extent, of island B.[2] It follows that as the number of individuals that decide to self-protect increases, the environmental degradation of island A (and thus the incentive to travel to island B) also increases. When the share of individuals who decide to go to island B becomes sufficiently high, the environmental quality of island A will be so low that everyone will want to move to island B with the result that island A's community, as a whole, will be in an undesirable situation since it no longer can enjoy free access to environmental goods, and must pay for access to private goods.

The strategic framework described above can be summarized as follows. Let us indicate the share of individuals that decide to go to island B as x and the share of those who remain in island A as $1-x$. We assume that the welfare of the agents that go to island B and of those that stay on island A can be represented by the following functions:

$$\Pi_B(x) = a - bx \qquad \Pi_A(x) = c - dx \tag{11.1}$$

where a, b, c, d are positive parameters satisfying the condition $d>b$. From this condition it follows that as x (and thus also environmental degradation) increases, the agents that self-protect by going to island B suffer a lower welfare loss than those who stay in island A.

In this context, we can see that if

$$a \geq c \tag{11.2}$$

holds, then the welfare from going to island B is always higher than that from staying on island A, whatever the share of agents x that decide to move to island B.

If, on the contrary

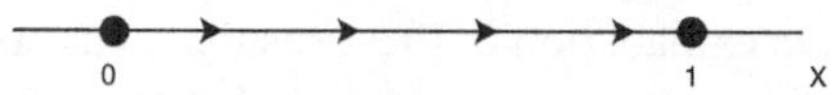

Figure 11.1 All individuals move to island B

Figure 11.2 All individuals remain on island A

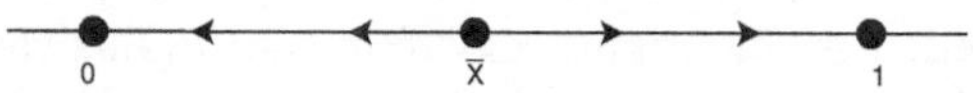

Figure 11.3 If the initial value of x is below (above) $\overline{x}$, then all individuals will stay on island A (move to island B)

$$a - b \leq c - d \tag{11.3}$$

then the opposite applies.[3]

Let us assume that there is a given learning mechanism underlying individual choice, which tends to spread the most remunerative strategy among the agents in the population (e.g. variable x evolves over time according to the so-called *replicator dynamics*; Weibull 1995). If so, in the two extreme cases described above, all individuals will end up making the same choice. In other words, variable x will either go to 1 (everyone moves to island B) or to zero (everyone stays on island A), whatever the initial share of agents x_0 who decide to move to island B (see Figures 11.1 and 11.2).

If the two extreme cases described above do not apply, namely, if

$$a < c \quad \text{and} \quad a - b > c - d \tag{11.4}$$

then there exists a value of x, $\overline{x}$, such that for every $x < \overline{x}$ the welfare of the individuals that move to island B is lower than that of the agents that stay on island A, whereas for every $x > \overline{x}$ the opposite is true.

In this case, if the initial share x_0 is less than the threshold level $\overline{x}$, then all individuals will stay on island A; and vice versa, if x_0 is greater than $\overline{x}$. If x_0 is equal to $\overline{x}$, then the two choices would provide the same welfare level and no individual would have any incentive to modify his/her choice (Figure 11.3).

Observe that if $x = 0$ (i.e., everyone stays in island A) the individual welfare will be $\Pi_A(0) = c$, while if $x = 1$ (i.e., everyone moves to island B) the individual welfare will be $\Pi_B(1) = a - b$, and $\Pi_A(0) > \Pi_B(1)$ if, and only if

$$a - b < c \tag{11.5}$$

By comparing condition (11.4) with conditions (11.1)–(11.3), we see that the agents' welfare is higher in $x = 0$ than in $x = 1$ whenever conditions (11.2) and (11.3) apply (Figures 11.2 and 11.3). Moreover, if b is high enough, the same result will occur in case (1) (Figure 11.1).[4]

To provide an intuitive explanation of these results let us consider, for instance, the well-known case of the Phi-Phi islands in Thailand. In 2000, one of these uncontaminated islands (Phi-Phi Leh) was the backdrop for the movie *The Beach*, which starred Leonardo di Caprio. Following the release of the movie, there was a dramatic increase in tourism to the Phi-Phi islands, which were among the most requested destinations of travel agencies around the world. However, the increase in the numbers of tourists and the local population (previously only a few fishermen and their families) led to the construction of many new buildings and the consequent depletion of the original environment.[5] In this case, therefore, if the number x of tourists that moves to Phi-Phi islands causes high environmental degradation of the islands (i.e., parameter b is high enough in our model), then moving to the Phi-Phi islands is no longer attractive and people are better off staying at home and enjoying their nearby beach rather than buying an expensive holiday to a famous beach that has been degraded and is overcrowded.

This simple analysis shows that the strategic selection process may lead agents to choose $x = 1$ although the welfare level in $x = 1$ is lower than in $x = 0$. Actually, no agent has an incentive to stay in island A if the other agents do not do the same. In this case, the choice of going to island B gives origin to an undesirable social convention, which represents a stable equilibrium of the economy. Any departure from this equilibrium, thus, requires external intervention in order to coordinate individuals' choices.

Policy implications: the role of public administration

This chapter has shown how individuals' attempts to protect themselves from environmental degradation may lead society into a private consumption trap characterized by high levels of self-protection consumption, high levels of environmental degradation and low levels of welfare. In order to escape this trap, which results from a coordination failure among agents, it is necessary for a sufficiently high share of individuals simultaneously to stop self-protecting. This has some implications for the environmental policies that public administrations should formulate to coordinate and modify agents' choices. In particular, in introducing ecological interventions, policy-makers should consider the opportunities available to individuals to enjoy their spare time for free, without the need for costly self-protection choices. This is particularly relevant for the proper projection and management of large urban centres, characterized by high population and production density, the environment where most self-protection choices occur and where the negative environmental externalities deriving from interactions among the agents are particularly evident. On the one hand urban centres offer a wide range of different ways to occupy one's leisure time; on the other hand almost none of these can be enjoyed for no cost. This damages relatively more the poorest families in the population, who cannot afford many of the costly self-protection activities available in towns.

Traffic bans over weekends are an example of public intervention that can contribute to reducing air and noise pollution in towns and inducing citizens to modify their consumption patterns. Forcing people to walk or cycle in city centres and simultaneously increasing the supply of free access places to citizens, may increase the environmental awareness of individuals, which, in turn, may promote greater public pressure on administrations for more environmental policies and better public transport.

Another important area of public intervention is the management of coastal areas. In this case, policy-makers should aim to reduce not only the environmental degradation of bathing areas close to large urban centres, but also the cost of access to such areas. It is important both to prevent these areas from further environmental degradation and to guarantee free access to beaches made private by the granting of concessions to private owners. Without such intervention, these (theoretically) free public goods will be transformed effectively into costly private goods.

In general, it would be desirable for the public administrations to identify all economic activities that may generate self-protection choices, and classify them according to multiple criteria, such as: (i) type of the subjects involved (individuals, firms, public institutions); (ii) location of the activity(ies); (iii) possible policy response to reduce or avoid the risk of private consumption traps.

Notes

1 Some empirical studies have tried to evaluate the dimension of the ecological self-protection choices (Leipert 1986; Leipert and Simonis 1988, 1989; Daly and Cobb 1990; Statistics Canada 1998; Escofet and Bravo-Peña 2007). While results differ according to the environmental defensive expenditures considered, there is wide agreement in the literature that these expenditures are not a negligible phenomenon and have been growing steadily over time.

2 The boats used to move from island A to island B, for instance, may end up polluting the sea surrounding island B. We comment further below on the relative impact of self-protection choices on the two islands.

3 Note that if condition (11.2) holds as equality and $x = 1$, then the agents are indifferent between staying in island A or moving to island B (i.e.,).

4 This result would occur *a fortiori* if we allowed b to grow above d, so that an increase in the defensive choices x damages island B relatively more than island A, thus reversing our initial assumption.

5 According to various environmental groups, some of these constructions did not have planning permission. The actual filming also provoked heated debate about the effects on the local environment. To avoid attracting international criticism and to enable filming to be concluded, Leonardo di Caprio promised the local authorities that the island would be restored to how it looked originally. The expenditure awarded by the film's producers to restoring the island when the film making finished (which can be considered private defensive expenditure), was so effective that some argue that the island ultimately looked better than it had done previously.

References

Antoci, A. and Bartolini, S. (1999) 'Negative Externalities as the Engine of Growth in an Evolutionary Context', FEEM Working paper 83, Milan: FEEM.

Antoci, A. and Bartolini, S. (2004) 'Negative externalities and labor input in an evolutionary game', *Environment and Development Economics* 9: 1–22.

Antoci, A., Borghesi, S. and Galeotti, M. (2008) 'Should we replace the environment? Limits of economic growth in the presence of self-protective choices', *International Journal of Social Economics*, 35(4): 283–97.

Bartolini, S. and Bonatti, L. (2002) 'Environmental and social degradation as the engine of economic growth', *Ecological Economics* 43: 1–16.

Bartolini, S. and Bonatti, L. (2003) 'Undesirable growth in a model with capital accumulation and environmental assets', *Environment and Development Economics*, 8: 11–30.

Daly, H.E. and Cobb, J.B. (1990) *For the Common Good*, Boston, MA: Beacon Press.

Escofet, A. and Bravo-Peña, L.C. (2007) 'Overcoming environmental deterioration through defensive expenditures: Field evidence from Bahía del Tóbari (Sonora, México) and implications for coastal impact assessment', *Journal of Environmental Management* 84(3): 266–73.

Hueting, R. (1980) *New Scarcity and Economic Growth. More Welfare Through Less Production?*, Amsterdam: Elsevier North Holland.

Leipert, C. (1986) 'Social costs of economic growth', *Journal of Economic Issues*, 20: 109–31.

Leipert, C. and Simonis, U.E. (1988) 'Environmental damage – environmental expenditures: Statistical evidence on the Federal Republic of Germany', *International Journal of Social Economics*, 15(7): 37–52.

Leipert, C. and Simonis, U.E. (1989) 'Environmental protection expenditures: The German example', *International Review of Economics and Business (RISEC)*, 36: 255–70.

Lopez, R. (2003) 'The policy roots of socioeconomic stagnation and environmental implosion: Latin America 1950–2000', *World Development*, 31(2): 259–80.

Statistics Canada (1998) *National Accounts and the Environment: Papers and Proceedings from a Conference*, Ottawa, 17–20 June 1997, Ottawa: Statistics Canada.

Weibull, J.W. (1995) *Evolutionary Game Theory*, Cambridge, MA: MIT Press.

12 Carbon Kuznets curves

Long-run dynamics and policy events

Massimiliano Mazzanti and Antonio Musolesi

Introduction

This chapter aims to contribute to the development of environmental Kuznets curve (EKC) research, which is not fully surveyed, given that the literature has overwhelmingly increased over the past years, in two main directions. First, we use modern econometric panel approaches capable of providing new evidence on EKC long-run dynamics. We employ recent homogeneous estimators – as those derived from panel cointegration analysis or those that explicitly take into account cross-section correlation – as well as heterogeneous estimators which allow individual slopes to be derived from sampling or Bayesian approaches. It is difficult *a priori* to decide between homogeneous and heterogeneous panel estimators. On the one hand, the increasing time dimension means that the slope homogeneity implicit in the use of a pooled estimator is questionable. On the other hand, most researchers agree about the use of homogeneous estimators since the efficiency gains from pooling often overcome their costs (Baltagi *et al.* 2000, 2002, 2004). Some researchers have suggested using intermediate estimators as Bayesian shrinkage estimators (Maddala *et al.* 1997) or the pooled mean group (PMG) estimator (Pesaran *et al.* 1999), allowing intercepts, short-run coefficients and error variances to differ freely across cross-sections, while long-run coefficients are held constant.

Secondly, we focus on a policy-relevant scenario, in which pro-Kyoto countries and the Umbrella Group, respectively led by the EU and the US, are compared in their EKC delinking performances. The main issue we address, looking at a long-run dynamics including the 1992–2001 pre- and post-Kyoto period, is why some countries were in favour of cutting CO_2 emissions and others opposed. A part of the underlying answer may be connected with the eventual historical decoupling of CO_2 and GDP, and 'policy-related' structural factors. More specifically, we may assume that the reason why some countries (EU, and within EU the northern countries including UK) supported Kyoto from the beginning and are supporting stricter targets (the 20-20-20 EU plan on energy and environmental efficiency) is that they took early actions in terms of economy restructuring and environmental policies. As early movers, they wanted to exploit the benefits related both to the Porter competitive advantages linked to new green technology markets (Porter and van der Linde 1995) and to the intrinsic advantage of reasoning in terms of CO_2 reductions decided in 1997 at Kyoto, which define a compliance with respect to 1990 levels. What happened between 1992 and 1997, and before 1992, matter(ed). Early movers could take advantage of Kyoto targets more than others on many perspectives.

Moreover, a lower elasticity and/or EKC evidence for a group (Northern EU) could explain stronger support for Kyoto, deriving from better historical environmental performance and favourable structural conditions. Nevertheless, this is the 'average picture'; reasoning at the margins, the current achievement of EKC shapes could be associated with higher marginal abatement costs, than reduced incentives for further efforts, if one excludes the objective of intensifying green and economic competitive advantages spurred by innovation investments (Jaffe *et al.* 1995; Mazzanti and Zoboli 2009). In terms of current policy negotiations, structural differences in EKC shapes could inform the allocation of burdens in the post-Kyoto phase (beyond 2012) for maximising economic efficiency at global level.

Though policy implications may be linked to the analysis of EKC paths, we believe that the literature has so far provided weak evidence. We shed light on such policy implications by taking quite a different angle from usual EKC analyses. In fact, the policy-oriented reasoning is key and derives both from: (i) a comparative assessment of EKC shapes for three group of countries, instead of analysing larger samples (OECD): the Umbrella Group, EU south and EU northern countries; and (ii) a series of structural break tests on the relevance, in affecting the CO_2 income time series, of exogenous political events, such as the 1992 Convention on climate change, the Kyoto Protocol, searching also for other sources of structural break. We bring together policy analysis and the study of advanced econometric methodologies, including intervention analysis aimed at highlighting how exogenous (policy) events affect a long-run structural dynamics. As recognised, policy events may be needed to reshape the business-as-usual EKC, by smoothing the bell and/or decreasing the income TP level. Economic growth matters for achieving higher environmental efficiency, but is not sufficient for sustainability. We provide some evidence on these issues.

Although the political agenda is changing, mainly in the US, our aim is to provide food for thought for political negotiators in the context of the post-Kyoto era, by examining the extent to which the structural differences of different groups of countries might explain their different policy perspectives and economic capabilities to tackle the climate change issue. Our empirical evidence provides useful information for both the current scenario, in which the US is slowly coming to recognise the need to tackle climate change, but favours flexible policy instruments, and the EU is leading Kyoto implementation (Kruger and Pizer 2004) and the post-Kyoto negotiation round, which should set the framework for the new climate change policy scenario.

The EKC: new applied perspectives

Recent works have highlighted, on the basis of newly updated data and new techniques, that there is some evidence supporting EKC shapes for CO_2, though different by geographical areas and by estimation techniques (Martinez-Zarzoso and Bengochea-Morancho 2004; Vollebergh *et al.* 2005; Cole 2003; Galeotti *et al.* 2006). Although the evidence is patchy, that is, heterogeneous across studies (which use different data with respect to time span and countries), there is some EKC evidence for CO_2 emerging for the OECD countries. This counterbalances other rather pessimistic views of no TP and a fragile EKC hypothesis (Harbaugh *et al.* 2002; Millimet *et al.* 2003).[1]

Thus, the evidence is far from conclusive and continues to grow based on ongoing research aimed mainly at verifying the robustness of results across different models. We would agree, therefore, that a best model for analysing EKC does not exist. In addition, at a certain level of advanced analysis, statistical fit comparisons across models are harder to implement. A consolidated assessment of results is still under way; critical points and heterogeneity of outcomes across models are currently being tackled. Finally, there is often a lack of policy relevance given the nature of data and the objectives of analyses. Here, we aim to check robustness across different models, in order to provide results that will inform policy. We briefly critique the more recent analyses, focusing on work that deals with dynamics and structural heterogeneity in panels.

There is a series of papers providing empirical evidence, which exploit flexible panel parametric specifications and non-parametric methods. Martinez-Zarzoso and Bengochea-Morancho (2004) analyse CO_2 data for the period 1975–1998, for 22 OECD countries, by applying the PMG estimator. The evidence favours an N-shape for the majority of OECD countries and an EKC inverted U-shape for the less-developed countries. The range of implied TPs is nevertheless too wide to lead to solid conclusions. Within the studies focusing on OECD, Cole (2005) applies the heterogeneous Swamy random coefficients estimator and concludes that the income–pollution relationship varies widely across countries. This suggests that the assumption of constants coefficients across countries in the traditional specifications is inappropriate.

Recent developments in the literature test the robustness of the EKC hypothesis using either flexible parametric specifications or partially or fully non-parametric models, or by looking at the cointegration properties of CO_2 time series; they have produced mixed results. Dijkgraaf and Vollebergh (2005) and Vollebergh *et al.* (2005) allow for both heterogeneity across countries and flexible (non-parametric) functional form, and show that traditional panel models with country-specific or country and time effects may present TPs within the observed income ranges; nevertheless, the null hypothesis of slope homogeneity is strongly rejected by the data. Dijkgraaf and Vollebergh analyse a sample of 24 OECD countries for 1960–1997. The most striking result is that time series analysis provides a different picture from heterogeneous panels. Only five out of 13 countries that show evidence of EKC dynamics present coherent results in the two frameworks.

Vollebergh *et al.* (2005) explore various parametrical and non-parametric specifications for a CO_2 dataset of OECD countries and find that EKC shapes are quite sensitive to the degree of heterogeneity in the panel estimations. Parametric models generate EKC shapes with quite low TPs, while the evidence is less robust from semi-parametric estimations. The non-parametric setting demonstrates the necessity to incorporate heterogeneity, which leads to the exploration of single country specific time series, and suggests caution in interpreting panel-based EKC outcomes if they do not in some way address the heterogeneity issue. The existence of an EKC curve in cross-country international frameworks, such as OECD country-based analyses, may depend on the balance between high-income countries showing an inverted U-shape dynamics and high-income countries that present a still-positive elasticity of emissions with respect to income. The role of semi-parametric and non-parametric EKC estimations is tackled by Azomahou *et al.* (2006), who use CO_2 data for 1960–1996 for 100 countries. They find that EKC shapes arise when a parametric panel model is used, but that a monotonic relationship emerges in both the non-parametric settings and the first difference regressions.

On the basis of recent theoretical and empirical developments, we maintain that EKC analysis is a useful tool for investigating income–environment relationships.

Model specification and data

Specification

Following the EKC and IPAT related literatures, and the main EKC oriented studies (Cole 2005; Stern 2004), the per capita CO_2 emissions from a country i in period t is modelled as a function of per capita GDP, with both variables expressed in logarithms:

$$y_{it} = f(x_{it}) \tag{12.1}$$

In order to allow for a non-linear/non-monotonic relation, we employ a (parametric) quadratic specification. Thus, the long-run environmental degradation–income relationship is given by:

$$y_{it} = \theta_{0i} + \theta_1 x_{it} + \theta_2 x_{it}^2 + \varepsilon_{it} \tag{12.2}$$
$$i = 1, \ldots, N, \quad t = 1, \ldots, T$$

where y_{it} is the logarithm of CO_2 emissions per capita, x_{it} is the logarithm of per capita GDP, α_i is individual effects and ε_{it} is the error term.

Similar to many other studies (Azomahou *et al.* 2006), we do not control for other possible determinants of CO_2 emissions, such as energy prices or technological change. Based on the present analysis, their investigation should be the subject of future research. There are several reasons for this specification. The first is data availability over long time series in terms of additional explanatory variables. Second, this specification allows for a greater comparability with existing studies. The third reason is more econometrically-oriented: although a specification that excludes other determinants of CO_2 emissions is not appropriate for measuring *ceteris paribus* the impact of GDP on CO_2 emissions, this kind of econometric specification is very useful for capturing the global effects of GDP on CO_2 including the indirect effects linked to the omitted variables which are correlated with GDP. Moreover, since we are not interested in obtaining the best prediction for CO_2, additional explanatory variables not correlated with GDP are irrelevant.

Data and samples

We focus on the developed regional areas that have been leading the climate change policy debate and were associated to Kyoto targets in 1997. We adopted the following samples composition: (a) Australia, Canada, Japan, New Zealand, Norway, the USA (the Umbrella Group); (b) Belgium, Denmark, Finland, France, Germany, Netherlands, Sweden, the UK (EU North); (c) Austria, Greece, Ireland, Italy, Portugal, Spain (EU South).[2]

Data on emissions are from the database on global, regional, and national fossil fuel CO_2 emissions prepared for the US Department of Energy's Carbon Dioxide Information Analysis Center (CDIAC). For our study, we use the subset of emissions data that matches the available time series on GDP per capita on the basis of joint availability, series continuity, and country definitions. This resulted in a sample of 109 countries for the period 1960–2001, from which we extracted the countries of interest for our study. Table 12.1 summarises the main variables used and the descriptive statistics.

Table 12.1 Descriptive statistics

	Mean	*Standard deviation*	*Min*	*Max*
Umbrella group				
CO_2 per capita	3.144921	1.393584	0.67	5.85
GDP per capita (GDPpc)	15,143.21	4,763.547	3,986.417	28,129.23
EU North				
CO_2 per capita	2.60875	0.5630643	0.91	3.88
GDP per capita (GDPpc)	14,203.73	3,759.392	6,230.359	23,160
EU South				
CO_2 per capita	1.488294	0.6085014	0.25	3.05
GDP per capita (GDPpc)	10,215.44	42,65.277	2,955.836	23,201.45

Notes: T= 1950–2001; CO_2 per capita in tons; GDP per capita in 1990 International 'Geary-Khamis' dollars

Panel estimators

The increased time dimension of panel data has generated new lines of research. A first strand of literature exploits panel data with time series procedures developed to deal with non-stationarity, spurious regression and cointegration (Kao and Chiang 2000; Phillips and Moon 1999). Another line of research developed both within and outside the framework of non-stationary panels concerns cross-section dependence (Bai and Ng 2004; Pesaran 2007; Driscoll and Kraay 1998). Finally, there is a third strand of literature that rejects the slope homogeneity implicit in the use of a pooled estimator, in favour of estimators allowing for individual slopes (Pesaran and Smith 1995; Hsiao *et al.* 1999).

We begin by assessing the (sensitivity of) results, and model performance, across the following *homogeneous estimators*: least square dummy (LSD) estimator allowing for individual fixed effects model (FEM), as basis; then the dynamic ordinary least squares (DOLS) estimator for the cointegrated panel data regressions (Kao and Chiang 2000; Saikkonen 1991); the PMG estimator proposed by Pesaran *et al.* (1999) which can be considered as an intermediate estimator since it allows intercepts, short-run coefficients and error variances to differ freely across cross-sections while holding long-run coefficients the same, the Driscoll–Kraay (DK) non-parametric estimator, which corrects the variance–covariance matrix for the presence of spatial as well as serial correlation and can be viewed as a variant of the Newey and West (1987) time series covariance matrix estimator; the GLS estimator of the seemingly unrelated regressions (SUR) specification proposed by Zellner (1962) allowing cross-section correlation via the individual error terms; and finally the dynamic SUR (DSUR) which takes account of cross-sectional correlation in a panel-cointegrated framework (Mark *et al.* 2004).

The first three estimators (FEM, DOLS, PMG) assume that all cross-section units are independent. In many cases, this assumption is clearly unrealistic from both economic and econometric points of view. First, the independence assumption is often at odds with economic theory. For instance, according to many economic models, agents tend to interact within and between cross-sections. Second, spatial dependence could also be the consequence of unobserved heterogeneity due principally to omitted observed common factors, spatial spillover effects, unobserved common factors, or general residual interdependence (Breitung and Pesaran, 2008). In these cases, standard techniques that do not take account of this dependence would yield inconsistent estimates of the parameter standard errors, producing incorrect inference and test statistics. Consequently, in order to correct for the presence of cross-sectional dependence, we also employ the last three estimators (DK, SUR, DSUR). We implement several tests of cross-section independence[3] and in all cases they strongly reject the null hypothesis that the errors are independent across countries.

All these estimators allow individual intercepts but common slopes. Although the increase in the time dimension allows us to reject the slope homogeneity implicit in pooled estimators, there are some features that render homogeneous estimators quite attractive. For example, Baltagi *et al.* (2000, 2002, 2004), find that homogeneous estimators have generally better forecasting[4] performance than their heterogeneous counterparts – mostly due to the simplicity, parsimony and stability of the parameter estimates.

Some authors suggest the use of heterogeneous estimators. For example, Baltagi *et al.* (2004) find that the superior forecasting performance of the homogeneous estimators is not a general result since both shrinkage estimators and the hierarchical Bayes estimator perform very well. There is another view that the use of heterogeneous estimators is related to the possible heterogeneity bias associated with the use of pooled estimators. As pointed out by Hsiao (2003), if the true model is characterised by heterogeneous intercepts and slopes, estimating a model with individual intercepts but common slopes could produce the false inference that the estimated relation is curvilinear. This situation corresponds to our empirical framework where: per capita GDP presents high variation across countries, the different groups of countries cannot be characterised by a common slope and, consequently, there is a high risk of estimating a false curvilinear relation when using homogeneous estimators.

Next, we apply the heterogeneous estimators. First, we apply the Swamy (1970) random coefficient GLS estimator, which is a weighted average of the individual least squares estimates where the weights are inversely proportional to their variance–covariance matrices. This is used as a benchmark. Then we apply and compare the mean group (MG) estimator proposed by Pesaran and Smith (1995) for dynamic random coefficient models. It is defined as the simple average of the OLS estimators relative to the individual equations expressed in ARDL; the hierarchical Bayes approach (Hsiao *et al.* 1999) which makes use of Markov chain Monte Carlo methods via Gibbs sampling. Hsiao *et al.* (1999) show that this is asymptotically equivalent to the MG estimator; the shrinkage estimators described in Maddala *et al.* (1997), that is, the empirical Bayes and the iterative empirical Bayes estimators. The parameter estimates are weighted averages (depending on the parameter variance–covariance matrices) of the pooled estimate and the individual time series estimates. Thus, the individual estimates are shrunk toward the pooled estimate.

Empirical evidence

We present evidence first comparing the long-run EKC dynamics of the three groups of countries, in order to highlight differences in shapes and eventual TP across different panel data models. Policy implications may derive from implicit considerations on the factors that differentiate the three groups. Secondly, in order to add an explicit policy flavour, we test through structural break analysis whether *policy events* such as the 1992 Climate Convention, which gave birth to Kyoto and the 1997 Kyoto Protocol itself, have affected the long-run dynamics. We also test the presence of other structural breaks affecting the emission–income relationship. We believe that the 1992 turning point may be even more relevant since it is a threshold that distinguishes between countries which began policy actions even in the period preceding the effective Kyoto Convention (and the country ratification) and countries that awaited Kyoto or beyond to take action.

EKC structural long-run dynamics

It should be noted that, while in some countries this inverted U-shaped pattern is symmetric, in others there is a non-symmetric pattern since the upward-bending portion of the curve does not swing back to the initial level of CO_2 per capita.[5]

Our evidence is mainly concerned with comparison of homogenous, and heterogeneous panel estimators, with the emphasis on cross-section correlation and adjustment dynamics.[6] As before, we examine six homogenous panel estimators (FEM as benchmark specification and DOLS,[7] PMG, DK,[8] SUR[9] and DSUR)[10] and heterogeneous-based estimators (Swamy, MG, empirical Bayes, iterative empirical Bayes, hierarchical Bayes). For each specification modelled, we examine the three samples of countries in terms of carbon–income shape (elasticity) and eventual TP, assessing whether this TP is within or outside the range of observed values. We present in Tables 12.2 and 12.3 only some of the results, for brevity. Mazzanti and Musolesi (2009) present evidence related to all above-mentioned estimators.

In relation to the first homogenous estimators, we note that the baseline FEM shows that quadratic specifications are significant for all the analysed cases, while the cubic specifications are not.[11] Nevertheless, the evidence is different across groups: while the TP for EU North is within the range of observed values ($13,000) this is not the case for the Umbrella Group and EU South, which show similar (slightly higher for EU South) TPs , around $45,000 to $50,000 per capita.[12]

Inverted U-shapes with a TP within the observed values for the EU North group and outside the observed values for the Umbrella and EU South groups, apply also to the other homogeneous estimators.

The DOLS estimator with two lags and no leads provides similar results to those from the FE specification and similar estimated TPs. DOLS assuming {(0, 1); (1, 1); (1, 2): (leads, lags)} show low estimate variability, with an estimated TP for EU North stable at around $11,000 and for the Umbrella and EU South groups, instead always outside the observed range of observations.

Introducing a certain degree of heterogeneity, as in the PMG estimator, does not modify the picture substantially. It should be noted, however, that while the TP estimate for pro-Kyoto countries is very stable across the different methods, allowing for cross-country correlation or for a limited degree of heterogeneity provides lower estimates of the quadratic specification and higher TPs for the umbrella group, well outside the range of observations.

To summarise the evidence from homogeneous specifications, apart from the differences in TP for the Umbrella and EU South groups, indicating non-existence of a robust EKC shape, the evidence for EU North is statistically and economically robust and is associated to EKC TPs in the range $11,000 to $14,000 per capita.

Comparison of the five heterogeneous panel data models presents slightly different evidence, which provides insights into economic and methodological perspectives.

The baseline specification is the well-known and extensively applied Swamy procedure which takes account of slope heterogeneity (Cole 2005).

Table 12.2 Homogenous estimators: FEM, DOLS, PMG

	Umbrella		*EU north*		*EU south*	
Group of countries	*coef.*	*t-stat.*	*coef.*	*t-stat.*	*coef.*	*t-stat.*
Model FEM						
GDPpc (linear)	3.716	7.146	16.888	14.762	2.862	8.493
GDPpc (quadratic)	–0.173	–6.407	–0.890	–14.833	–0.132	–7.333
EKC shape	inverted U		inverted U		inverted U	
Turning point ($1995)	46,160.715		13,195.623		51,067.782	
Turning point range	out		in		out	
Model DOLS						
GDPpc (linear)	6.948	6.010	13.606	6.069	1.701	2.343
GDPpc (quadratic)	–0.316	–5.092	–0.731	–6.130	–0.081	–1.985
EKC shape	inverted U		inverted U		inverted U	
Turning point ($1995)	57,894.784		10,990.809		38,163.230	
Turning point range	out		in		out	
Model PMG						
GDPpc (linear)	3.041	2.067	12.846	5.375	3.117	4.485
GDPpc (quadratic)	–0.126	–1.64	–0.687	–5.452	–0.152	–4.000
EKC shape	inverted U		inverted U		inverted U	
Turning point ($1995)	174,113.091		11,491.294		28,375.730	
Turning point range	out		in		out	

The specification does not drastically modify the evidence presented above, but reveals other factors. For example, both the Umbrella and EU South groups, which showed an EKC shape with TPs outside the range, are now consistent; they do not present bell shapes, but demonstrate linear relationships between income and CO_2 emissions. Elasticity is slightly lower than 0.5, which is a sign of *relative delinking* in the dynamics of these countries, at least based on the group average. EU North shows evidence of absolute delinking, with a TP of around $13,000.

The different evidence is worth noting since it highlights that all homogenous panel estimators, although in our case not showing robust EKC shapes from an economic point of view, tend to erroneously capture output as a non-linear path. This quadratic trend may be the result of our not taking account of heterogeneity in income–environment relationships for certain groups of countries. Checking for outliers or 'non-average' situations could modify the picture in homogeneous settings. However, in focusing our analysis on structural heterogeneity, we are

Table 12.3 Heterogeneous estimators: Swamy, MG, Hierarchical Bayes

Group of countries	*Umbrella*		*EU north*		*EU south*	
	coef.	*t-stat.*	*coef.*	*t-stat.*	*coef.*	*t-stat.*
Model Swamy						
GDPpc (linear)	0.473	4.778	17.492	4.135	0.464	6.705
GDPpc (quadratic)	…	…	–0.922	–4.229	…	…
EKC shape	monotonic		inverted U		monotonic	
Turning point ($1995)			13,172.68			
Turning point range			in			
Model MG						
GDPpc (linear)	0.475	3.006	12.262	4.966	0.436	4.955
GDPpc (quadratic)	…	…	–0.654	–5.070	…	…
EKC shape	monotonic		inverted U		monotonic	
Turning point ($1995)			11,785.41			
Turning point range			in			
Model Hierarchical Bayes						
GDPpc (linear)	3.600	36.327	17.494	201.080	2.178	25.326
GDPpc (quadratic)	–0.163	–3.630	–0.922	–36.888	–0.088	–2.667
EKC shape	inverted U		inverted U		inverted U	
Turning point ($1995)	62,501.4		13,159.87		236,806.82	
Turning point range	out		in		out	

Note: (…) means not included given not significance

providing an average picture of single country's stories for income–environment dynamics.

In terms of other heterogeneous-based estimators, we note that the outcomes of the MG model and the Swamy procedure are very similar.

For the Bayesian approaches, we focus on hierarchical Bayes estimators. Mazzanti and Musolesi (2009) present other additional Bayesian estimators.

Hierarchical Bayes is the only situation when an EKC emerges, but for anti-Kyoto countries, the quadratic terms are very low and the estimated TPs are well above the range of observed values. Instead, the TP for EU North is fairly consistent with the TPs in homogeneous models, showing again coherency across models as far as EU North countries are concerned.

To sum up, the set of heterogeneous-based estimators, Bayesian or not, provide robust evidence of an EKC for the EU North countries and only relative delinking for the other two groups. We note that the consistency of estimates across models (level of the coefficients) is stronger for heterogeneous models, which present lower variability. Also, they show that tackling heterogeneity using specific tools provides a clearer understanding of the income–environment relationship, although we can also highlight that the differences for the group showing EKC shapes (EU North) are slight even when comparing homogeneous and heterogeneous models. Overall, then, our evidence is very robust. We can be confident that the shapes and TPs we estimated are representative of the real phenomenon.

Evaluating policy events in the climate change international arena

We now assess the impact of a *postulated policy event* on the carbon–income relationship. The *intervention analysis* developed by Box and Tiao (1975) is the methodology of reference. Some previous studies have successfully applied this methodology in modelling the economic effect of public policies.[13]

In order to test the hypothesis that the occurrence of the 1992 UN Framework Convention[14] and the consequential 1997 Kyoto Protocol has modified the relation between emissions and economic development, the following model is specified:

$$y_t = f(\mathbf{x}_t, \boldsymbol{\theta}) + g(\boldsymbol{\delta}, \boldsymbol{\omega}, \boldsymbol{\psi}, t) \tag{12.3}$$

where y_t denotes per capita CO_2 emission, $f(\mathbf{x}_t, \boldsymbol{\theta}) = \theta_0 + \theta_1 x_t + \theta_2 x_t^2 + \varepsilon_{ut}$ corresponds to the EKC relation previously estimated, where x_t is per capita GDP. Finally, $g(\boldsymbol{\delta}, \boldsymbol{\omega}, \boldsymbol{\psi}, t)$ allows for some deterministic effects of time t, the effects of some exogenous variables, $\boldsymbol{\psi}$, measured through the vectors of parameters $\boldsymbol{\delta}$ and $\boldsymbol{\omega}$. This can be modelled combining a step function with an exponential (or first order) transfer function:

$$\psi_t = \text{Step\_1993}_t = \begin{cases} 1, & \text{if } t \geq 1993 \\ 0, & \text{otherwise} \end{cases} \tag{12.4}$$

$$g(\boldsymbol{\delta}, \boldsymbol{\omega}, \boldsymbol{\psi}, t) = \frac{\omega}{1 - \delta B} \psi_t$$

where B is the backward shift operator such that $B^i y_t = y_{t-i}$ .The magnitude of the impact that occurred after 1992 (alternatively 1997) is given by ω_1 and δ_1 is the rate of decay of the variation. When $\delta<1$ the series will reach a new steady state and the steady state gain is $\omega/(1-\delta)$, while when $\delta = 1$, a step change in the input produces a ramp function in the output. Finally, $\delta>1$ will produce an exponential pattern decay. Alternatively, a gradual effect can be modelled using a ramp function:

$$\psi_t = \text{Ramp\_1993}_t = \begin{cases} t - 1992, & \text{if } t \geq 1993 \\ 0, & \text{otherwise} \end{cases} \tag{12.5}$$

$$g(\lambda, \boldsymbol{\psi}, t) = \lambda \psi_t$$

where λ measures the magnitude of the change in the trend of the series. In that follows the ML estimation results are provided for both specifications and

Table 12.4 Structural analyses on Policy events

Specification	θ_0	θ_1	θ_2	ω	δ	λ	*TC8386*	$\omega 80$	$\delta 80$	*AIC*	*SBC*
UMBRELLA											
Step_1993	–79.33(0.00)	16.39(0.00)	–0.83(0.00)	0.013(0.01)	1.02(0.00)		–0.06(0.00)			–209.26	–198.98
Ramp_1993	–74.91(0.00)	15.47(0.00)	–0.78(0.00)			0.013(0.00)	–0.06(0.00)			–212.03	–203.34
Step_1997	–71.65(0.00)	14.75(0.00)	–0.75(0.00)	0.04(0.01)	0.57(0.02)		–0.07(0.00)			–201.86	–191.57
Ramp_1997	–66.66(0.00)	13.70(0.00)	–0.69(0.00)			0.018(0.00)	–0.07(0.00)			–204.88	–196.19
EU_SOUTH											
Step_1993	–49.95(0.00)	10.29(0.00)	–0.52(0.00)	0.015(0.03)	1.02(0.00)					–179.53	–170.97
Ramp_1993	–47.99(0.00)	9.86(0.00)	–0.50(0.00)			0.014(0.00)				–185.75	–178.80
Step_1997	–45.00(0.00)	9.18(0.00)	–0.46(0.00)	0.041(0.04)	0.61(0.04)					–174.28	–165.72
Ramp_1997	–43.00(0.00)	8.73(0.00)	–0.43(0.00)			0.020(0.00)				–179.72	–172.78
EU_NORTH											
Step_1993	–5.15(0.85)	0.64(0.91)	0.002(0.99)	–0.01(0.25)	1.22(0.00)			–0.09(0.00)	0.74(0.00)	–140.51	–128.51
Ramp_1993	–5.29(0.84)	0.67(0.90)	0.0002(0.99)			–0.020(0.01)		–0.10(0.00)	0.72(0.00)	–139.69	–129.40
Step_1997	–5.31(0.84)	0.68(0.90)	0.003(0.99)	–0.05(0.06)	0.81(0.00)			–0.09(0.00)	0.77(0.00)	–142.73	–130.74
Ramp_1997	–5.28(0.84)	0.67(0.90)	0.0002(0.99)			–0.03(0.00)		–0.09(0.00)	0.77(0.00)	–144.07	–133.79

Note: p-values in brackets

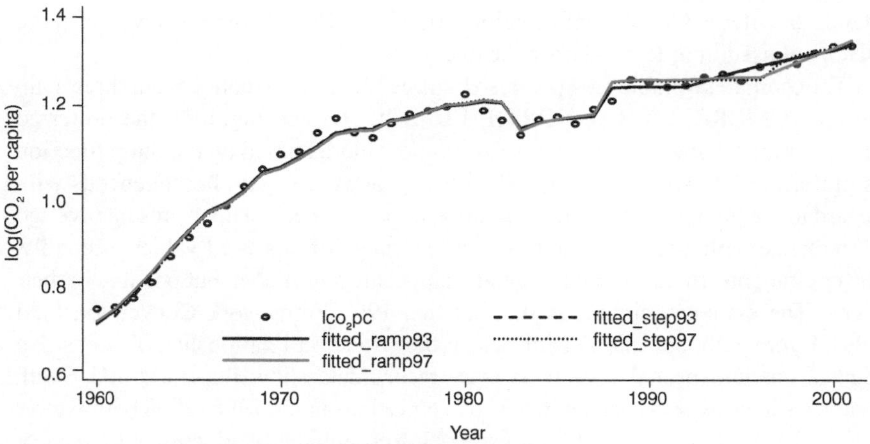

Figure 12.1 Real and fitted values, Umbrella Group

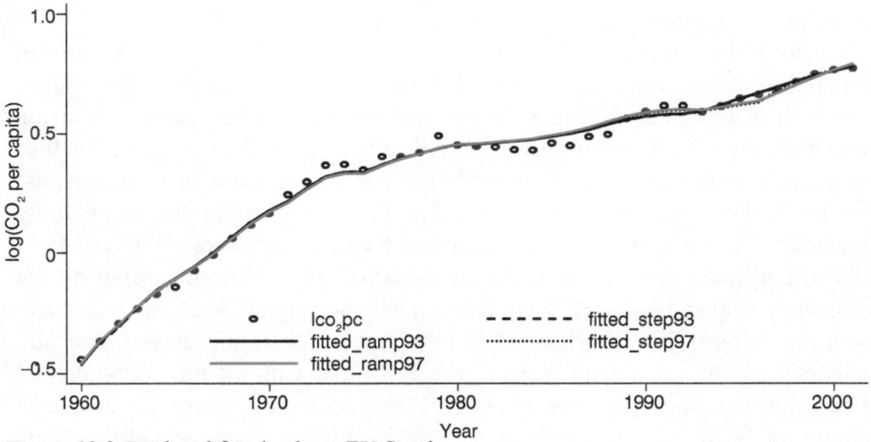

Figure 12.2 Real and fitted values, EU South

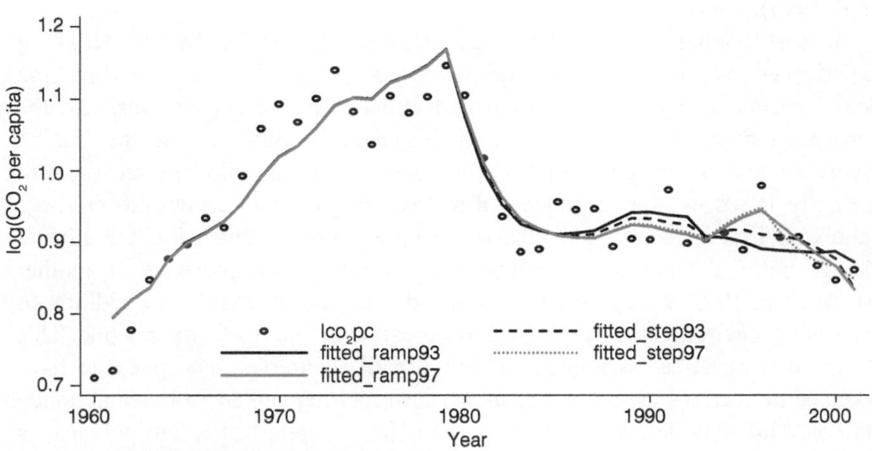

Figure 12.3 Real and fitted values, EU North

standard criteria (Akaike information criteria, AIC; Schwartze–Bayes criteria, SBC) are used in order to choose the one preferred.

The estimates in Table 12.4 (see also Figures 12.1 to 12.3) concern our three main groups (UMBRELLA, EU_NORTH, EU_SOUTH) and highlight the preferred specifications that we comment on. Overall, the model based on the ramp function is preferred. The Umbrella[15] and EU South groups are again homogeneous with regard to the income–environment relationship: first, EKC shapes present inverted U evidence with a turning point outside the range of observed values, secondly, the coefficients, representing the trend change, are significant, but positive in their signs. The evidence highlights the fact that 1992 Framework Convention (and 1997 Kyoto Protocol) did not impact on the structural relationship. The positive sign is not unexpected insofar as even recent data show that most EU South countries have experienced an increase in emissions in the 1998–2008 period after Kyoto (EEA 2008) and are still far from being compliant to reduction with respect to 1990 levels. The high growth (and low oil price) period that followed 1992 did not witness a significant break in the income–emission elasticity.

As far as EU North countries are concerned, a negative coefficient instead emerges in association to the trend change, with high statistical significance. Instead, it is worth noting that, following the application of an outliers selection procedure, a permanent shift also occurred in the early 1980s (namely 1980 as key year). This has been modelled combining a step function with an exponential transfer function similar to eq. (4). Table 12.4 presents for EU North only regressions including the climate change and Kyoto conventions (1992 and 1997 breaks) and the aforementioned change occurred after 1980, measured by the parameters $\omega 80$ and $\delta 80$, indicating respectively the magnitude of the impact and its decay pattern. The statistical 1980 break can refer to and be economically explained by the second oil shock (namely 1979), with all the consequential effects on the post-recession (1981–82) restructuration phase of advanced economies, beginning around the early 1980s, which is characterised by efforts towards higher energy efficiency and increasing environmental innovations (Jaffe *et al.* 1995).

At least looking at the picture until 2001, it seems that the absolute delinking experienced by Northern EU countries is attributable both to path-breaking policy events such as the environmental climate change conventions, and the consequential Kyoto Protocol, and to exogenous events such as the Iranian revolution and associated second oil price shock with the following recession of the early 1980s, with a consequential restructuring of such economies on more energy/environmental efficiency basis. Along a temporal dimension, the climate change political emphasis emerging in the 1990s, in the presence of another recession in 1992–93, could partly descend from the oil shocks, in addition to increasing environmental awareness coherent with the EKC framework. This evidence is nevertheless limited to Northern EU countries that appear to have taken earlier actions in terms of economy restructuring and environmental policy actions. This may be a key reason for their strong support of Kyoto policies, as most (innovative and composition effects related) efforts were already in place in 1997. Lagging or anti-Kyoto countries face(d) larger investments regarding CO_2 reduction, though probably lower marginal costs of abatement. Both issues matter for the current political/negotiation agenda. The current economic crisis may change the political agenda towards green investments, though we note that contrary to the exogenous break we highlighted, it is characterised by low oil prices which do not incentive environmental efficiency investments. Other motivations have to be found.

Conclusions

This study has provided new EKC evidence based on long panel data series and exploiting various advanced panel estimators. We focus attention on three groups of countries in the political economy arena related to Kyoto (and post-Kyoto) frameworks: the Umbrella Group led by the US; the EU North group, which is the most proactive in climate change issues; and the EU South group of countries, which have lower incomes per capita and generally lower-level commitment to climate change. Our results are relevant from both an economic and a methodological point of view, and are useful for informing the post-Kyoto negotiation rounds.

We find that the Umbrella and EU South groups, which are less in favour of stringent climate policies, have not experienced a carbon Kuznets curve yet, as expected, that is absolute delinking, although there is evidence of relative delinking in the carbon–income relationship, with elasticities around 0.45–0.50. The EU North countries show robust EKC shapes across panel specifications. It should be noted that both homogeneous and heterogeneous panel models provide similar evidence, with minor differences across models. However, the latter seem to perform better, insofar as they capture the real shape of Umbrella and EU South countries' relationship, which is linear and not bell shaped with TPs outside the range. When structural heterogeneity is relevant homogeneous panels inevitably capture the average trends for the groups of countries considered, and are more likely to give bell-shaped relationships for output.

Tests on the relevance of exogenous policy events, such as the 1992 climate change convention and the 1997 Kyoto Protocol provide further and complementary policy-oriented evidence. Exogenous path-breaking 'policy events' appear to matter. The income–emission relationship is in fact affected by such events, at least for Northern EU countries that present EKC shapes over the period. The post-1992–93 period, characterised by high growth and low oil prices, was a preliminary arena where some countries took early actions in environmental/energy policy aimed at increasing the GDP efficiency. Scandinavian countries implemented green fiscal reforms aimed at achieving economic–environmental double dividends; the Netherlands and Germany introduced some elements of ecological taxation in the system; the UK concluded the restructuring of the economy towards services, away from manufacturing. All such interventions added up to the post oil shock energy efficiency restructuring already in place, and may be part of the drivers of the evidence we find.

We indeed find some signs that the absolute delinking associated to EU northern countries may largely depend on exogenous shocks that occurred well before the environmental conventions of the early 1990s; some results suggest in fact that the early 1980s oil price shock and recession, and following energy-economy restructuring on a more environmental efficient basis, was probably a major event in determining a turning point. This group of countries took advantage of the oil shock to restructure the economy and took early actions for setting a green technological competitive advantage. These pre-Kyoto facts largely explain their strong commitment towards climate change, as they are better positioned and already on the track. Given the sunk costs of investments, economies of scale and complementarity between green and standard innovation investments, such countries could lead the post-Kyoto phase as well, after being mostly compliant with 1997 Kyoto targets. The lower costs of abatement compared with those faced by the USA for Kyoto compliance (see Chapter 10 of this book) may be determined not by marginal cost considerations, but by the amount of investments, partly sunk, that such countries have already put in place, positioning on a current greener track, which may or may not be maintained over time. The new climate change orientation of the US administration under Obama is a possible shock event of the US and the climate change arena as a whole.

The reason for their higher commitment to Kyoto principles lies in the (social and policy) choice to acknowledge the opportunities presented by climate change 'markets' (green products, environmental innovation) as a basis for new competitive advantage, based on the production of an (impure) public good such as carbon abatement, combined with economic gains for the economy. Being an early mover in the market may enable these advantages to be consolidated in the medium long run.

Nevertheless, it is worth noting that considerations of global economic efficiency should also put the weight of future abatement on advanced countries that have not reached a TP in the income–environment relationship and are not compliant. On average, these lagging countries have more scope for incremental efforts towards abatement of carbon emissions (among other figures, the current consumption of oil is around 26 barrels per capita in the US and 12 in the average EU country, thus even lower in some northern EU countries; on a total-energy perspective respective figures are 60 versus 30 barrels per capita), and then presumably lower marginal costs under usual assumptions on abatement-cost functions and technological conditions. Climate change negotiation and policy initiatives in future years will demonstrate whether countries currently lagging in terms of delinking and commitment to climate change policy, will be able to combine carbon abatement and the achievement of environmental (innovation and policy) competitive advantages to become the basis for a race to the top of the ranking, not, as a possible opposite scenario, a divergence in emission/income trends.

Notes

1 Wagner (2006), Muller Furstenberger and Wagner (2007), and Galeotti *et al.* (2006) are rather sceptical about EKC and test the robustness of the EKC hypothesis, analysing CO_2 series.
2 Note that the groups are homogeneous in terms of policy perspectives on climate changes. Some Umbrella countries have finally ratified the Kyoto protocol, which nevertheless is only the first step to addressing climate change at global level. The EU countries have all ratified the protocol, and now have different views on the post-Kyoto phase and on the EU objectives of reducing emissions by 20 per cent by 2020, a target led by EU North.
3 The Lagrange multiplier approach of Breusch and Pagan (1980), the CD test of Pesaran (2004) and the Frees (1995, 2004) statistics.
4 Forecasting-oriented studies include Auffhammer and Carson (2008) and Schmalensee *et al.* (1998) among others.
5 We present the results obtained using panel data regression approaches. They show the advantages of capturing the indirect effects linked to the omitted variables correlated with GDP. Moreover, they show some comparative advantage with respect to non-parametric panel approaches such as are used by Azomahou *et al.* (2006). On the one hand, a non-parametric panel approach allows for a free functional form; on the other, the parametric methods we employ allow for slope heterogeneity, adjustment dynamics and cross-sectional correlation which are at least as relevant as functional form.
6 Auffhammer and Carson (2008).
7 Implemented specifying a zero lead and two lags.
8 The maximum lag considered in the autocorrelation structure (L) is set equal to one. Alternative values of L (2, 3, 4) provide similar strand error estimates.
9 Constraining the slope coefficients to be equal across equations while allowing for different intercepts.
10 As for DOLS, it was implemented specifying zero leads, two lags, and individual FE. Also, in this case, results do not differ if leads and lags change.
11 Here, and subsequently, cubic specifications (terms) are never statistically significant, as expected.
12 The maximum value for income per capita is $28,129 per capita for the Umbrella Group, $23,160 for EU North and $23,201 for EU South.
13 Sharma and Khare (1999), who assess the effectiveness of CO_2 pollution control legislation in India; Fomby and Hayes (1990), who examine the impact of redistributive policies in the US. Other relevant contributions include Lloyd *et al.* (1998), Murray *et al.* (1993) and Thompson and Noordewier (1992) who evaluate respectively anti-cartel policies, anti-drinking campaigns and incentive programmes on automobile sales.
14 This postulated break is coherent with the hypothesis that some countries may have acted as early movers with regard to the Kyoto arena post-1997, on the basis of either/both the 1992 convention or/and even before 1992 events.
15 A temporary change relative to the period 1983–86 (TC8386) has been detected and introduced in the model (Table 12.4).

References

Aldy, J. (2006) 'Per capita carbon dioxide emissions: convergence or divergence?', *Environmental and Resource Economics*, 33: 533–55.

Auffhammer, M. and Carson, T. (2008), 'Forecasting the path of China's CO_2 emissions using province-level information', *Journal of Environmental Economics and Management*, 55: 229–47.

Azomahou, T., Laisney, F. and Phu Ngayen, V. (2006), 'Economic development and CO_2 emissions: a non parametric panel approach', *Journal of Public Economics*, 90: 1347–63.

Bai, J. and Ng, S. (2004), 'A panic attack on unit roots and cointegration', *Econometrica*, 72: 1127–77.

Baltagi, B.H., Griffin, J.M. and Xiong, W. (2000), 'To pool or not to pool: Homogeneous versus heterogeneous estimators applied to cigarette demand', *Review of Economics and Statistics,* 82: 117–26.

Baltagi, B.H., Bresson, G. and Pirotte, A. (2002), 'Comparison of forecast performance for homogeneous, heterogeneous and shrinkage estimators: some empirical evidence from US electricity and natural-gas consumption', *Economics Letters,* 76: 375–82.

Baltagi, B.H., Bresson, G. and Pirotte, A. (2004), 'Tobin q: forecast performance for hierarchical Bayes, shrinkage, heterogeneous and homogeneous panel data estimators', *Empirical Economics,* 29: 107–13.

Box, G.E.P. and Tiao, G.C. (1975), 'Intervention analysis with applications to economic and environmental problems', *Journal of the American Statistical Association*, 70: 70–92.

Breitung, J. and Pesaran, M. H. (2008), 'Unit roots andcCointegration in panels'. In L. Matyas, and P. Sevestre (eds), *The Econometrics of Panel Data,* third edition. Cheltenham: Edward Elgar.

Breusch, T.S. and Pagan, A.R. (1980), 'The Lagrange multiplier test and its application to model specifications in econometrics', *Review of Economic Studies,* 47: 239–53.

Cole, M.A. (2003), 'Development, trade and the environment: How robust is the environmental Kuznets curve?' *Environment and Development Economics,* 8: 557–80.

Cole, M.A. (2005), 'Re-examining the pollution–income relationship: a random coefficients approach', *Economics Bulletin*, 14: 1–7.

Copeland, B.R. and Taylor, M.S. (2004), 'Trade, growth and the environment', *Journal of Economic Literature*, 42: 7–71.

Dijkgraaf, E. and Vollebergh, H. (2005), 'A test for parameter homogeneity in CO_2 panel EKC estimations', *Environmental and Resource Economics*, 32: 229–39.

Driscoll, J.C. and Kraay, A.C. (1998), 'Consistent covariance matrix estimation with spatially dependent panel data', *Review of Economics and Statistics*, 80: 549–60.

EEA (2008), *Greenhouse Gas Emission Trends and Projections in Europe 2008*, Copenhagen: European Environment Agency.

Fomby, T.B. and Hayes, K.J. (1990), 'An intervention analysis of the war on poverty: poverty's persistence and political-business cycle implications', *Journal of Econometrics,* 43: 197–212.

Frees, E.W. (1995), 'Assessing cross-sectional correlation in panel data', *Journal of Econometrics,* 69: 393–414.

Frees, E.W. (2004), *Longitudinal and Panel Data: Analysis and Applications in the Social Sciences*, Cambridge: Cambridge University Press.

Furstenberger, G. and Wagner, M. (2007), 'Exploring the EKC hypothesis: theoretical and econometric problems', *Ecological Economics,* 67: 648–60.

Galeotti, M., Lanza, A. and Pauli, F. (2006), 'Reassessing the environmental Kuznets curve for CO_2 emissions: a robustness exercise', *Ecological Economics,* 57: 152–63.

Harbaugh, W., Levinson, A. and Wilson, D. (2002), 'Re-examining the empirical evidence for an environmental Kuznets curve', *Review of Economics and Statistics*, 115: 1167–99.

Hsiao, C. (2003), *Analysis of Panel Data,* Cambridge: Cambridge University Press.

Hsiao, C., Pesaran, M.H. and Tahmiscioglu, A.K. (1999) 'Bayes estimation of short-run coefficients in dynamic panel data models'. In: Hsiao, C., Lahiri, K., Lee, L.F. and Pesaran, M.H. (eds), *Analysis of Panels and Limited Dependent Variable Model,* Cambridge: Cambridge University Press, pp.268–96.

Kao, C. and Chiang, M.H. (2000), 'On the estimation and inference of a cointegrated regression in panel data', *Advances in Econometrics*, 15: 179–222.

Kruger, J. and Pizer, W.A. (2004) 'The EU Emissions Trading Directive: Opportunities and potential pitfalls', Discussion paper 04-24, Washington DC: Resources for the Future.

Lloyd, T., Morrissey, O. and Reed, G. (1998), 'Estimating the impact of anti-dumping and anti-cartel actions using intervention analysis', *Economic Journal*, 108: 458–76.

Jaffe, A., Peterson, S.R., Portney, P.R. and Stavins, R. (1995), 'Environmental regulation and the competitiveness of US manufacturing: what does the evidence tell us?, *Journal of Economic Literature*, 33: 132–63.

Maddala, G.S., Trost, R.P., Li, H., Joutz, F. (1997), 'Estimation of short-run and long-run elasticities of energy demand from panel data using shrinkage estimator', *Journal of Business and Economic Statistics,* 15: 90–100.

Mark, N.C., Ogaki, M. and Sul, D. (2004), 'Dynamic seemingly unrelated cointegration regression', *Review of Economic Studies,* 72: 797–820.

Martinez-Zarzoso, I., Bengochea-Morancho, A. and Morales Lage, R. (2007), 'The impact of population on CO_2 emissions: evidence from European countries', *Environmental and Resource Economics*, 38: 497–512.

Mazzanti, M. and Musolesi, A. (2009), 'Carbon Kuznets curves: long-run structural dynamics, Kyoto coalitions and policy events', FEEM Working paper 87, Milan: FEEM.

Mazzanti, M. and Zoboli, R. (2009), 'Environmental efficiency and labour productivity: trade-off or joint dynamics?', *Ecological Economics*, 68: 1182–94.

Millimet, D., List, J. and Stengos, T. (2003), 'The EKC: real progress or misspecified models?' *Review of Economics and Statistics*, 85: 1038–47.

Murray, J.P., Stam, A. and Lastovicka, J.L. (1993), 'Evaluating an anti-drinking and driving advertising campaign with a sample survey and time series intervention analysis', *Journal of the American Statistical Association*, 88: 50–6.

Musolesi, A., Mazzanti, M. and Zoboli, R. (2009), 'A Bayesian approach to the estimation of EKC for CO_2', *Applied Economics*, i-first.

Newey, W.K. and West, K.D. (1987), 'A simple, positive semi-definite, heteroskedasticity and autocorrelation consistent covariance matrix', *Econometrica*, 55: 703–8.

Pesaran, M.H. (2004), 'General diagnostic tests for cross-section dependence in panels', Cambridge Working Papers in Economics 0435, Faculty of Economics, Cambridge: University of Cambridge.

Pesaran, M.H. (2007), 'A simple panel unit root test in the presence of cross-section dependence', *Journal of Applied Econometrics*, 22: 265–312.

Pesaran, M.H. and Smith, R. (1995), 'Estimating long-run relationships from dynamic heterogeneous panels', *Journal of Econometrics*, 68: 79–113.

Pesaran, M.H., Shin, Y. and Smith, R. (1999), 'Pooled mean group estimation of dynamic heterogeneous panels', *Journal of the American Statistical Association*, 94: 621–34.

Phillips, P.C.B. and Moon, H.R. (1999), 'Linear regression limit theory for non-stationary panel data', *Econometrica*, 67: 1057–111.

Porter, M. and Van der Linde, C. (1995), 'Toward a new conception of the environment–competitiveness relationship', *Journal of Economic Perspectives*, 9: 97–118.

Saikkonen, P. (1991), 'Asymptotic efficient estimation of cointegration regressions', *Econometric Theory*, 7: 1–21.

Schmalensee, R., Stoker, T. and Judson, R. (1998), 'World carbon dioxide emissions 1950–2050', *Review of Economics and Statistics*, 80: 15–27.

Sharma, P. and Khare, M. (1999), 'Application of intervention analysis for assessing the effectiveness of CO pollution control legislation in India', *Transportation Research Part D*, 4: 427–32.

Stern, D.I. (2004), 'The rise and fall of the environmental Kuznets curve', *World Development,* 32: 1419–38.

Swamy, P. (1970), 'Efficient inference in a random coefficient regression model', *Econometrica*, 38: 311–32.

Thompson, P.A. and Noordewier, T. (1992), 'Estimating the effects of consumer incentive programs on domestic automobile sales', *Journal of Business & Economic Statistics*, 10: 409–17.

Vollebergh, H. and Dijkgraaf, E. (2005), 'A test for parameter homogeneity in CO_2 panel EKC estimations', *Environmental and Resource Economics*, 32: 229–39.

Wagner, M. (2006), 'The Carbon Kuznets Curve: a Cloudy Picture Emitted by Bad Econometrics?', Working Paper Economic Series 197, Wien: Institute for Advanced Studies.

Zellner, A. (1962), 'An efficient method of estimating seemingly unrelated regressions and test for aggregation bias', *Journal of the American Statistical Association*, 57: 348–68.

13 The European Union Emissions Trading Scheme

CO_2 price drivers during the learning experience

Emilie Alberola, Julien Chevallier and Benoît Chèze

Introduction

The primary objective of the European Union Emissions Trading Scheme (EU ETS) is to regulate carbon dioxide (CO_2) emissions from the most energy-intensive industry sectors. Installations have to surrender as many allocated allowances as their previous year's emissions. By allowing the emergence of a price, or European Union Allowance (EUA), for each ton of CO_2 emitted, the EU ETS is aimed at sending price signals to industry operators. The latter can select a combination of capital investments, operating practices and emissions releases to minimize their total abatements costs and allowance expenses. The main characteristic of such cap-and-trade systems is that firms have the freedom to decide on their own abatement strategies, based on their risk preferences and perceptions of regulatory uncertainties, and available technical opportunities (i.e. in terms of fuel-switching).

The EU ETS became functional in January 2005. Its creation required the deployment of new governance processes involving the European Commission (EC), member state authorities, and operators in the more than 10,000 installations covered by Directive 2003/87/CE. The development of technical tools was required to deal with the delivery of industry allowances and to guarantee the monitoring of transactions.

This fast-track operational implementation suffered from some delays (e.g., the start-up of the Polish registry), but no major setbacks. In 2005, the EU carbon finance sector was born. In its first three years, the players active in the market multiplied, organized exchanges were established, and trading activity developed rapidly. The volume of over-the-counter and in-the-market trade in emissions allowances has continued to grow. This chapter analyses the price fundamentals of CO_2 allowances during 2005–2007.

Key design issues on the EU ETS

This section briefly reviews the main design features of the European carbon market. The EU ETS was created by Directive 2003/87/CE. Across its 27 member states, the EU ETS covers large plants in CO_2-intensive emitting industrial sectors with a rated thermal input exceeding 20 MWh. One allowance exchanged on the EU ETS corresponds to one ton of CO_2 released into the atmosphere. During Phase I (2005–2007), 2.2 billion allowances per year were distributed; 2.08 billion allowances per year will be distributed during Phase II (2008–2012). With a value of around €20 per allowance, the launch of the EU ETS thus corresponds to a net creation of wealth of around €40 bn. In January 2008, the EC extended the scope of the EU trading system to include other sectors, such as aviation and petro-chemicals, by 2013, and confirmed its functioning for a third Phase until 2020. Below we provide price details for CO_2 allowances.

EUA price drivers

First, we define the emissions shortfall factor; second, we detail the allowance supply and demand mechanisms; third, we describe the influence of policy decisions, other energy markets and weather events; fourth, we investigate the likely effects of economic growth on the development of carbon prices.

The emissions shortfall factor

Whereas in energy markets the central issue related to price is the state of stocks, in the EU ETS it is the expected emissions shortfall factor in each compliance year that is important. The emissions shortfall factor, defined as the difference between verified emissions and allocated allowances within a given compliance year, depends on the emissions abatements required by the cap, which are unknown, but are estimable based on available data.

Supply and demand mechanisms in allowances

As is the case in other commodity markets, the EUA is driven by the balance between supply and demand, and by other factors related to market structure and institutional policies. However, EUAs exhibit strong features of a non-standard commodity: installations do not need physically to hold allowances in order to produce; they need only to match them to verified emissions in their annual compliance report to the EC.

On the supply side, the first market period was quite limited, as no other carbon credits than Phase I EU allowances could be used for compliance. Neither Kyoto credits, even if transactions started in 2003, nor Phase II EUAs (through borrowing) were allowed. Thus, the number of allowances available on the market was confined strictly to the number of allowances initially allocated to installations and to new entrants. Note, though, that the equilibrium market price can be affected by flexibility mechanisms, such as banking and borrowing provisions and the inflow of project mechanism credits. However, these features were not applicable during Phase I.

While the number of allowances is fixed by each member state government through the National Allocation Plans (NAPs), their demand is function of expected CO_2 emissions. In the short and long runs, the level of emissions depends on a large number of factors.

Short-term abatement decisions and, thus, demand for allowances, are mainly driven by unexpected fluctuations in energy demand, energy prices and weather conditions (temperatures, rainfall and wind speed). CO_2 emissions in the EU ETS are linked one-to-one to fossil fuel (e.g. oil, gas, coal) use. In turn, the demand for fossil fuels depends on their absolute and relative prices. The marginal fuel switching costs from high carbon-intensive sources of energy to low carbon-intensive sources for power and heat generation are the most important measures for carbon abatement in the short run, since power and heat operators are major actors in the EU ETS.[1]

Other factors influencing the demand for allowances are unexpected fluctuations in short-run demand for energy. As outlined by Alberola *et al.* (2008), CO_2 prices are affected by unexpected climatic variations, including temperature, rainfall and wind. Cold winters increase the need for heating using electricity or other fuel and hot summers lead to higher demand for electricity for cooling and lower rate of utilization of nuclear power plants due to reduced cooling by rivers. Rainfall, wind speed and hours of sunshine affect the share of power generated by carbon-free heat generation from hydropower, wind and solar energy. Overall, the weather is widely acknowledged to play an important role in explaining CO_2 prices.

In the long run, the demand for allowances can be affected by economic growth and marginal abatement costs. Apart from very long-term investments, such as carbon capture and storage (CCS), end-of-pipe technologies cannot reduce CO_2 emissions. Long-term marginal abatement costs are determined by investment in low carbon-intensive energy utilities and energy efficiency measures. Overall, with respect to high investment costs and uncertainties about the future of the EU ETS in 2005, in deciding about their allowances, firms were in a situation of *short-term* abatement costs during the first trading period.

Institutional factors, energy markets and climate

Market sentiment and its influence on EUA demand are among the most difficult factors to capture. Market sentiment refers to such factors as uncertainty about future energy prices and policy decisions, which are very important in terms of investors' expectations and the formation of their risk strategies.

Below, we briefly review an emerging literature that analyses the empirical relationships between EU carbon prices and their determinants during 2005–2007. Institutional decisions on the overall stringency of the cap have an impact on carbon price formation through initial allocation. Also, any decision or announcement from the regulator (member state or EC) may induce changes in the market players' behaviour. EC official communications are essential to achieve the best information flow on the net short/long positions of installations[2] in Phase I. Disclosure of compliance results may be seen as the basis of changes in market participants' expectation and strategies. In April 2006, when member states and the EC disclosed the verified emissions for all EU ETS installations in 2005, the gap between the initial allocations for industry and business-as-usual emissions forecasts appeared positive, and caused a sharp fall in carbon prices of all maturities (more than 50 per cent in four days).

Political choices can also impact on the functioning of the market: in October 2006, the EC announced plans for stricter validation of NAPs during Phase II which reinforced the general feeling that the market could be short in 2008–2012. After October 2006, Phase II prices increased and stabilized at over €20/ton of CO_2. Alberola *et al.* (2008) identify statistically two structural endogenous breaks in the spot price time-series for 2005–2007: the first compliance period break goes from 24 April 2006 to 12 June 2006; the second break based on stricter EC Phase II validation, is from 26 October 2006 onwards.[3]

It should also be noted that energy prices are the most important short-term drivers of demand for EUAs, due to the ability of power generators to switch between fuel inputs (Mansanet-Bataller *et al.* 2007; Delarue and D'haeseleer 2007; Alberola *et al.* 2008; Bunn and Fezzi 2007; Ellerman and Feilhauer 2008). As Alberola *et al.* (2008) show, errors in energy price forecasting drove EUA price changes during 2005–2007, but their influence changed with the occurrence of institutional events. Brent, natural gas, coal, clean dark and spark spreads, as well as switch prices[4] all impact significantly on carbon spot prices, but their influence varied following the two structural breaks in the time series, and the corresponding sub-periods under consideration.

As already mentioned, weather conditions impact on EUA prices by influencing energy demand. During winter 2006, colder than decennial average temperatures had the expected statistical positive impact on EUA price changes. In summer 2006 and winter 2007, higher than decennial average temperatures negatively affected carbon price changes (Alberola *et al.* 2008).

The puzzle of economic activity effects on CO_2 and prices

Economic activity is perhaps the most obvious and least understood driver of CO_2 price changes. Economic growth leads to increased energy demand and higher industrial production in general. The first empirical analysis of these issues was Alberola *et al.* (2009a).

Figure 13.1 shows that economic activity in Europe has been relatively robust since the launch of the EU ETS in 2005: according to Eurostat, Gross Domestic Product (GDP) in the EU 25 grew by 1.9 per cent in 2005, and 3.0 per cent in 2006. Industrial production, seasonally adjusted by Eurostat, rose by 2.8 per cent in 2005, and 4.4 per cent in 2006. There was a similarly rising trend in 2007.

Figure 13.2 shows that production growth in the EU ETS sectors differed during the first three years of the scheme. In 2005, four sectors (coke, refineries, glass and ceramics) recorded negative growth at the aggregate EU level. The EU

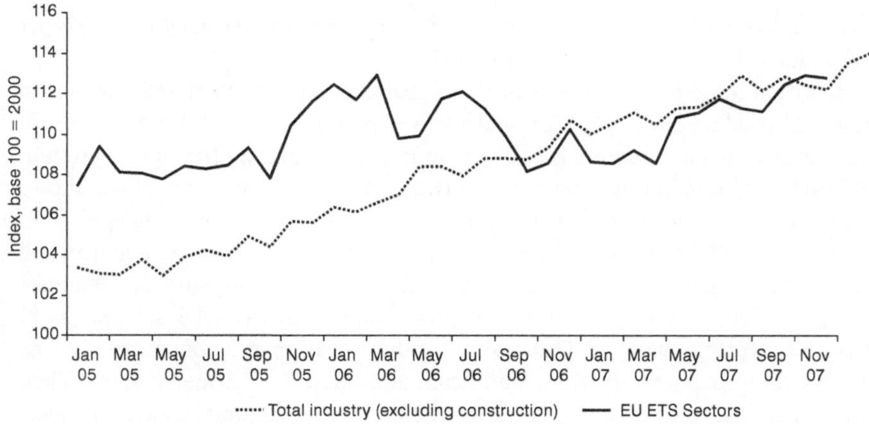

Figure 13.1 Industrial production in EU 27 during 2005–2007
Source: Eurostat– calculations from the Mission Climat, Caisse des Dépôts.

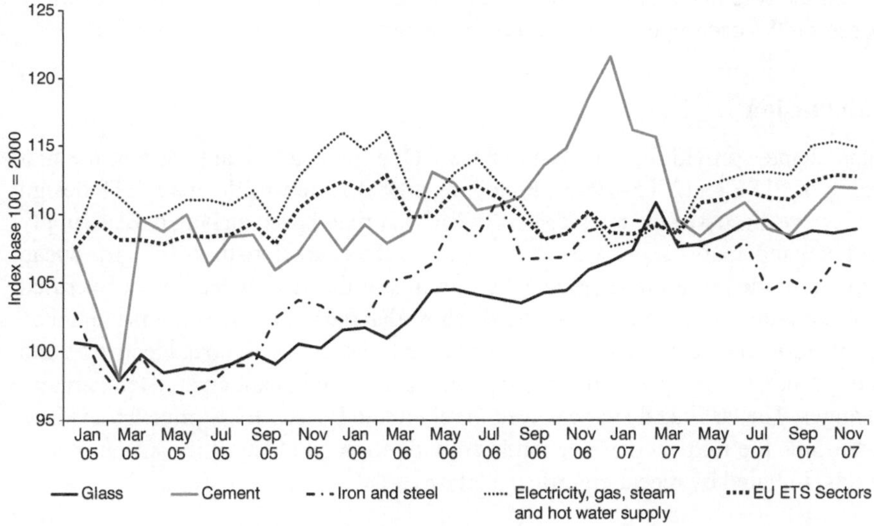

Figure 13.2 Industrial production in EU ETS sectors
Source: Eurostat– calculations from the Mission Climat, Caisse des Dépôts.

coke sector decreased by 20 per cent, mainly localized in Poland (–32 per cent) and the UK (–7 per cent). Annual glass production decreased strongly by 9.88 per cent in Germany, 4.28 per cent in the UK, 5.02 per cent in Spain, and 13.72 per cent in Poland. Production of ceramics decreased in 2005 in all member states except Poland where it increased by 19.55 per cent. Five sectors (paper, cement, iron, metal and combustion) recorded positive production growth in 2005,[5] with combustion increasing in all member states, including by 2.24 per cent in the UK and 12.95 per cent in Italy.

In 2006, whereas industrial production increased in seven industries, the combustion sector recorded a decrease of 4.93 per cent at EU level, and in all countries from 1.83 per cent in Spain to 8.13 per cent in Germany and 9.36 per cent in Italy. Production stabilized in 2007.

Based on these characteristics of the global economic context and changes in industrial production in EU ETS sectors, Alberola *et al.*'s (2009a) econometric exercise is aimed at disentangling potential impacts – ranging from production to environmental conditions – on carbon prices. Their econometric analysis shows that three out of nine sectors had a significant effect on EUA price changes from July 2005 to end April 2007. The three sectors are combustion, paper and iron and account for 78 per cent of the allowances allocated. They identify the channels through which variations in industrial production from EU ETS sectors might affect EUA price changes. They examine the variation in production and the net short/long positions that are significant and affect CO_2 price changes. They demonstrate the role played by yearly compliance positions and production peaks in this new market. In Alberola *et al.* (2009b) the authors extend their results to a country-specific analysis of industry sectors for 2005–2007. These later results reveal the role played by Germany's power sector on overall EUA price changes based on the decomposition of industrial activity and CO_2 emissions effects.

Conclusion

This chapter provides an analysis of the CO_2 price determinants during the first phase of EU ETS (2005–2007). Following a brief review of the main ETS design issues, we discussed the main drivers of carbon prices based on both classic supply and demand analysis, but also with respect to the characteristics of the European cap-and-trade programme (especially in terms of the aggregated level of verified emissions and compliance). Finally, we show that the main price drivers are linked to institutional decisions, other energy market prices (oil, gas, coal, electricity, and more generally fuel-switching in the power sector), and unanticipated temperature changes. The last puzzle to be solved in relation to Phase I EUA price drivers is to determine the influence of economic activity, through changes in CO_2 emissions levels, initiated by recent empirical analyses.

Notes

1 The main sector, i.e. with the largest in terms of emissions covered by the EU ETS in Phase I, is the power sector, which accounts for over 50 per cent of emissions capped by the scheme. Its emissions abatement costs are assumed to be the lowest compared with other sectors, based on switching from coal to gas fuel.

2 An installation is defined as short (long) when it records a deficit (surplus) of allowances allocated with respect to actual emissions (Ellerman and Buchner 2008).

3 Phase I spot prices decreased towards zero up to the end of Phase I due to banking restrictions implemented between 2007 and 2008 (Alberola and Chevallier 2009).

4 The dark spread is the theoretical profit that a coal-fired power plant makes from selling a unit of electricity, having purchased the fuel required to produce that unit of electricity. The spark spread refers to the equivalent for natural gas-fired power plants. The equilibrium between these clean spreads represents the carbon price above which it becomes profitable for an electric power producer to switch from coal to natural

gas, and below which it is beneficial to switch from natural gas to coal. As long as the market carbon price is below this switching price, coal plants are more profitable than gas plants – even after taking account of carbon costs. This switching price is more sensitive to changes in natural gas prices than coal prices (Kanen 2006). These three profitability indicators are used to determine the preferred fuel for power generation.

5 The combustion sector accounts for the largest share of installations in the EU ETS with 70 per cent of the EU allocation. The combustion sector is defined in a different way by each member state and contains numerous sub-activities. Based on CITL data and the NACE Classification Rev.1 C–F, Trotignon and McGuiness (2007) and Trotignon *et al.* (2008) distinguish between large electricity production plants, district heating facilities (cogeneration where details are available) and other installations.

References

Alberola, E. and Chevallier, J. (2009) 'European carbon prices and banking restrictions: Evidence from Phase I (2005–2007)', *Energy Journal*, 30(3): 107–36.

Alberola, E., Chevallier, J. and Chèze, B. (2008) 'Price drivers and structural breaks in European carbon prices 2005–2007', *Energy Policy*, 36(2): 787–97.

Alberola, E., Chevallier, J. and Chèze, B. (2009a) 'The EU Emissions Trading Scheme: Disentangling the effects of industrial production and CO_2 emissions on carbon prices', *International Economics*, 116: 93–126.

Alberola, E., Chevallier, J. and Chèze, B. (2009b) 'Emissions compliances and carbon prices under the EU trading scheme: A country specific analysis of industrial sectors', *Journal of Policy Modeling*, 31: 446–62.

Bunn, D. and Fezzi, C. (2007) 'Interaction of European Carbon Trading and Energy Prices', FEEM Working paper 63, Milan: FEEM.

Delarue, E.D. and D'haeseleer, W.D. (2007) 'Price determination of ETS allowances through the switching level of coal and gas in the power sector', *International Journal of Energy Research*, 31: 1001–15.

Ellerman, A.D. and Buchner, B.K. (2008) 'Over-allocation or abatement? A preliminary analysis of the EU ETS based on the 2005–06 emissions data', *Environmental and Resource Economics*, 41: 267–87.

Ellerman, A.D. and Feilhauer, S. (2008) 'A Top-down and Bottom-up Look at Emissions Abatement in Germany in Response to the EU ETS', *MIT CEEPR Working Paper*, 2008–017.

Kanen, J.L.M. (2006) *Carbon Trading and Pricing,* Environmental Finance Publications, London.

Mansanet-Bataller, M., Pardo, A. and Valor, E. (2007) 'CO_2 prices, energy and weather', *Energy Journal*, 28(3): 67–86.

Trotignon, R. and McGuiness, M. (2007) 'Technical Memorandum on Analysis of the EU ETS Using the Community Independent Transaction Log', *MIT-CEEPR Working Paper*, 2007–012.

Trotignon, R., McGuiness, M. and Delbosc, A. (2008) 'European CO_2 Market and the CITL: The Trial Period Under Scrutiny', Research Report Mission Climate, Caisse des Dépôts, Paris.

Index